Workbook of

Second Edition

ATMOSPHERIC DISPERSION ESTIMATES

An Introduction to Dispersion Modeling

with floppy diskette

D. Bruce Turner

Trinity Consultants, Inc.
Chapel Hill, North Carolina

LEWIS PUBLISHERS
Boca Raton   Ann Arbor   London   Tokyo

11586893

Learning Resources
Centre

Library of Congress Cataloging-in-Publication Data

Turner, D. Bruce, 1931–
 Workbook of atmospsheric dispersion estimates : an introduction to
dispersion modeling / D. Bruce Turner. — 2nd ed.
     p.   cm.
  Includes bibliographical references and index.
  ISBN 1-56670-023-X
  1. Air—Pollution—Mathematical models.   2. Smoke plumes-
-Mathematical models.   3. Atmospheric diffusion—Mathematical
models.   I. Title.
 TD883.T86   1994
 628.5′3′015118—dc20   93-43585
                                                                CIP

© 1994 by CRC Press, Inc.
Lewis Publishers is an imprint of CRC Press

No claim to original U.S. Government works
International Standard Book Number 1-56670-023-X
Library of Congress Card Number 93-43585
Printed in the United States of America   4 5 6 7 8 9 0
Printed on acid-free paper

Dedication

This workbook is dedicated to the memory of Larry Niemeyer, who guided and encouraged many air pollution meteorologists. I will always remember his direction and counsel.

Preface

This workbook is a revision to the workbook of the same title first published in 1967. Although Gaussian modeling has limitations, much of the air quality simulation modeling being done today to estimate impact of proposed sources in response to regulatory requirements is Gaussian. The intent of the workbook is 1) to provide an introduction to the bases for dispersion calculations and to dispersion modeling to those that need a starting point in this subject area, and 2) to provide a ready reference to dispersion equations and example calculations for air quality professionals already familiar with dispersion.

Acknowledgments

The author appreciates the encouragement of Robert A. McCormick, who suggested that the original workbook be compiled. The assistance of Francis Pooler is acknowledged. The encouragement of my wife, Ann, during the evenings and weekends while preparing this manuscript is also appreciated.

About the author

Bruce Turner is a Certified Consulting Meteorologist and serves as senior consultant at the Chapel Hill, North Carolina office of Trinity Consultants, Inc. He has a B.A. in mathematics from Carleton College, was trained as an Air Force Weather Officer at the University of Chicago and has an M.S. in meteorology from the University of Michigan. He retired in 1989 from the National Oceanic and Atmospheric Administration's laboratory while on assignment to the Environmental Protection Agency, where he was involved in air quality simulation model development, model evaluation, and making models available to users. He received the Air and Waste Management Association's Frank A. Chambers award on June 25, 1992. This award is given "for outstanding technical achievement, universally recognized as a major contribution to the science and art of air pollution control." He was cited for his leadership "in practical application of meteorology to air pollution problems, especially for his outstanding work with atmospheric dispersion estimates."

Contents

Chapter 3.   Effective Height of Emission

Chapter 4.   Special Topics

Chapter 5. **Putting Gaussian Methods Into Perspective**

Chapter 6. **Using Computers for Dispersion Estimates**

Chapter 7. **Use of the Computer Diskette**

Chapter 8. **Example Problems**

References

Appendix

Floppy Diskette

One 3 1/2" floppy diskette: READ.ME file gives instructions for use.

Contains two executable programs for instructional (not production) use.

CHAPTER 1.

INTRODUCTION

How Does the Atmosphere Disperse Material and What Is a Dispersion Model?

The air motions in the atmosphere transport pollutants that are released in the atmosphere. The smaller time and space scales of motion serve to disperse pollutants in the air by mixing these pollutants with air having lower pollution levels and thus lowering the air pollutant concentrations with time after release from a particular source.

1.1 Averaging Periods for Data

It is convenient to consider averages of meteorological parameters over one-hour time periods. This is primarily because the system of meteorological observations that are taken world-wide at airports to support aircraft operations are taken at one-hour intervals. These are usually archived so that they are later available for other purposes including their use for air pollution meteorology. These hourly airport observations generally do not represent averages over the hour but instead are spot observations that are taken approximately during the ten minutes prior to the clock hour. In situations where on-site data are collected for air pollution purposes, the parameter values are typically averaged over one-hour time periods, but may be averaged and archived over shorter time periods.

1.2 Wind

Wind is a velocity, a vector quantity having a direction and speed. Although the wind vector can occur in three dimensions, it is common only to consider the horizontal components of the wind. By convention the wind direction is the direction from which the wind comes. For the airport observations the wind is reported with a resolution of 10 degrees. North is 0 degrees or 360 degrees, and east is 90 degrees. The horizontal wind is motion in response to both the horizontal pressure gradient and the horizontal temperature gradient. For further information on the causes of wind, see an introductory text on meteorology.

The wind speed units of airport observations is usually in knots, nautical miles per hour. However for most air pollution purposes the wind is considered in meters per second, $m\ s^{-1}$.

1.2.1   Effect of Direction

The effect of wind direction is to determine the direction of transport of released pollutants. Because wind direction is the direction from which the wind blows, a west wind would cause pollution to move toward the east from the source.

1.2.2 Effect of Speed

For pollution releases that are continuous the pollutants are diluted right at the point of release, such as at the stack top. The resulting concentrations in the plume are inversely proportional to the wind speed; double the wind speed and the concentration in the plume becomes half of that with the lighter wind.

Because there is a frictional effect on the wind by the ground's surface and the roughness elements on the surface, the wind slows at heights close to the ground. A typical variation of wind speed with height during daytime and at night is shown in Figure 1.1. Typical variations of temperature with height are also shown. The resulting slowing of the wind near the ground or increase of the wind with height away from the ground is frequently approximated mathematically with the power-law wind profile.

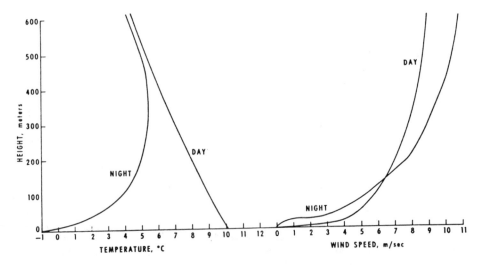

Figure 1.1 Examples of variation of temperature and wind speed with height (after Smith, 1963).

The power-law is:

$$u_z = u_a (z/z_a)^p \qquad (1.1)$$

where
u_z	wind speed, m s^{-1}, at vertical height z above ground
u_a	wind speed, m s^{-1}, at anemometer height
z	vertical height above ground, m
z_a	anemometer height above ground, m
p	exponent dependent primarily on atmospheric stability which varies from around 0.07 for unstable conditions and to about 0.55 for stable conditions.

1.3 Effect of Surface Roughness

The objects on the surface (roughness elements) over which the wind is flowing will have a frictional effect upon the wind speed nearest the surface. This effect is depicted in Figure 1.2 for three different surfaces. The numbers given at various heights are the wind speed at that height relative to the gradient wind, in percent. The gradient wind occurs at the height above the surface where the effects of the surface are no longer felt. It can be considered as the free-stream flow and is in response to the pressure and temperature gradients. Both the height and the spacing of the roughness elements on the surface will influence the frictional effect on the wind. A single parameter, the surface roughness length, z_0, is used to signify this effect. Typical values are given in Table 1.1.

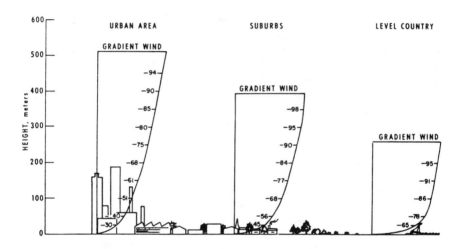

Figure 1.2 Examples of variation of wind with height over different size roughness elements (figures are percentages of gradient wind) (from Davenport, 1963).

Table 1.1. Surface roughness length, z_0, for typical surfaces, m.

Urban	1.0-3.0
Coniferous Forest	1.3
Deciduous Forest (summer)	1.3
Deciduous Forest (winter)	0.5
Desert Shrubland	0.3
Swamp	0.2
Cultivated Land (summer)	0.2
Cultivated Land (winter)	0.01
Grassland (summer)	0.1
Grassland ((winter)	0.001
Water	0.0001

1.4 Turbulence

1.4.1   What Is Turbulence?

Turbulence is essentially the motions of the wind over the time scales smaller than the averaging time used to determine the mean wind. Turbulence consists of circular whirls or eddies of all possible orientations, horizontal eddies, vertical eddies and all orientations in between. These turbulent eddies serve to disperse pollutants by mixing with air having lower pollutant concentrations. The causes of these eddies or whirls are primarily due to mechanical or buoyant generation of turbulence.

1.4.2   Mechanical Turbulence

Wind moving past vegetation or structures creates mechanical turbulence. The stronger the wind the greater the degree of mechanical turbulence generated. Also, the larger the roughness elements (structures, vegetation) on the surface, the greater the mechanical turbulence. Mechanical turbulence can also be created by wind shear, a slower moving air stream next to a faster moving current. This occurs in the vertical as wind is slowed near the earth's surface by the friction of the surface. At higher heights above the ground the wind speeds are higher. The shearing action between the variation of these two airstreams produces mechanical turbulence. The generation of mechanical turbulence is always positive.

1.4.3   Buoyant Generation of Turbulence

The heating or cooling of air near the earth's surface generates buoyant turbulence. At one extreme, during mid-day with clear skies and light winds, the heating of the sun creates an upward heat flux at the ground surface and this heats the air in the lower layers. With extreme heating, large convective eddies generate upward-rising thermals that may extend vertically on the order of 1000 to 1500 meters. This represents the generation of positive buoyant turbulence at its greatest.

At the other extreme, at night with light winds, the outgoing infrared radiation cools the ground and the air adjacent to it resulting in a downward heat flux at the surface. This cooling near the ground while the temperature of the air above remains relatively unchanged results in the creation of a temperature inversion in the layer near the ground. (An inversion is a vertical temperature structure that is inverted from the usual decrease of temperature with height.) The influence of the inversion causes the atmosphere to stabilize and resist vertical motions. This reduction of vertical exchange reduces vertical dispersion. Dispersion is the spreading of material, such as pollutants released into the atmosphere. Under this condition negative buoyancy is being generated. This negative buoyancy will even tend to damp out some of the mechanical turbulence.

1.5 Relation of Turbulence to Vertical Thermal Structure

The existence of thermals, mentioned above, represents the extreme of instability, or unstable atmospheric conditions. Since in making dispersion estimates we are usually dealing with time periods of one hour, several thermals will move with the wind past any fixed position. This will result in a series of upward motions during the thermals and slower descending motions in the compensating downward flow between the thermals. During an hourly period these series of upward and downward vertical motions will result in extreme vertical dispersion.

At night, with clear skies and light wind, a minimum of turbulence will exist. Much of the small amount of mechanical turbulence that is generated will be damped out by the negative buoyant turbulence caused by the stable thermal structure (inversion). The resulting eddies of relatively small scale will cause only a minimum amount of dispersion.

Between the above two extremes is the condition characterized as neutral. For this condition the net heat flux at the ground is near zero, so there is little or no heating or cooling of the ground and the air adjacent to it. The vertical thermal structure is a slight decrease of temperature with height. (As one goes upward in the atmosphere there is less atmosphere above you and hence less pressure due to the decrease in weight. This allows a parcel of air moving from a lower height to a higher height to expand due to this decrease in pressure. This expansion results in cooling. This rate of cooling is called the adiabatic lapse rate and is approximately 0.0098 K m^{-1}. Adiabatic is without adding or subtracting heat. A parcel forced downward heats at the same rate.) With net heat flux near the ground equal to zero, the temperature variation with height is near the adiabatic rate. These atmospherically neutral conditions can be caused by 1) cloudy conditions which inhibit incoming or outgoing radiation, 2) windy conditions that will rapidly mix vertically any heating or cooling that is attempted at the surface, or 3) transitional situations near sunrise and sunset when the atmosphere is changing from stable to unstable or vice versa. With neutral atmospheric conditions an intermediate level of dispersion takes place. Table 1.2 summarizes these characteristics.

Table 1.2. Characteristics of Atmospheric Conditions.

Atmospheric Condition	Typical Conditions	Heat Flux	Thermal Structure	Nature of Turbulence
Unstable	Mid-Day Clear Sky Light Wind	Net Upward	Super-Adiabatic	Considerable Horizontal and Vertical
Neutral	Windy or Cloudy or Transition	Zero	Near Dry Adiabatic	Mid-Range
Stable	Nighttime Clear Sky Light Wind	Net Downward	Near Isothermal or Inversion	Damps Out Vertical

1.6 Appearance of Continuous Visible Emissions
to Vertical Thermal Structure - Plume Types

Unstable - Under unstable conditions a visible continuously released plume will appear to exist as large loops, thus the description "looping." See Figure 1.3. Although the appearance would infer a roller coaster type of motion, the plume appearance results from portions of the plume being caught in the upward motions of the thermals as they move past the point of release and continuing that upward motion as the thermal is transported downwind. At a somewhat later time the descending air surrounding the thermals moves past the point of release and that part of the pollutant plume is caught in downward moving air and continues to descend as it moves away from the point of release. By considering the motions that occur at a single distance downwind as various pieces of the plume move by, the result is a considerable amount of vertical motion, both upward and downward, over the period of an hour. The result is considerable vertical dispersion of the effluent over this period. Since the eddy structures causing these motions have orientations in all possible directions there is considerable horizontal dispersion also taking place over averaged one-hour periods.

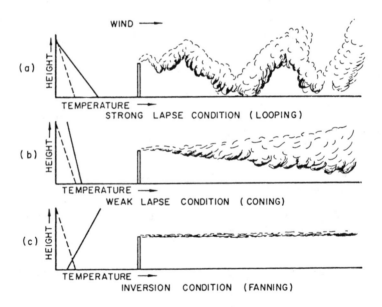

Figure 1.3   Vertical dispersion of continuous plumes related to vertical temperature structure. Source: Hewson (1964).

Neutral -   Under neutral conditions most of the turbulence is usually mechanical in nature rather than buoyant. The turbulent eddies have many different orientations and the resulting dispersion is relatively symmetrical in the horizontal and vertical resulting in the appearance of the plume as a cone with its vertex at the source. Thus the term "coning" to describe the condition. Because of a dominance of mid-range eddy sizes causing this condition, the time-averaged plume will not differ too much from the instantaneous appearance.

Stable -   Under stable conditions the thermal structure inhibits vertical motion so there is almost no vertical dispersion.   The thermal structure does not inhibit horizontal motions so there may be a considerable variety of horizontal plume appearances under stable conditions, from a very narrow ribbon to a considerable amount of horizontal spreading. This latter appearance is "fanning."

1.7 What Is a Dispersion Model?

What is a dispersion model? It is possible to solve a single equation for plume rise to estimate the effective height of the plume from a single source.   Then by use of a dispersion equation, an estimate of the concentration from this source can be made for a single receptor.   Can the solution of these two equations be considered a dispersion model? It is generally not considered to be a model. It is usually when a computer is used for the repetitious solution of these equations that we refer to the calculations as a dispersion model.

The primary inputs to a dispersion model, Figure 1.4, consist of emission information, meteorological data, and receptor information. For regulatory approved models such as ISC (EPA, 1987a), the required emission information consists of the coordinates for the location of the source, the physical stack height, the inside stack top diameter, the stack exit velocity, and the stack temperature. The meteorological parameters that are required hourly for input to this model are Pasquill stability class, wind direction, wind speed, temperature, and mixing height. These inputs are entries to the dispersion model, which is a mathematical simulation of the chemistry and physics of the atmosphere.   The outputs from the model are primarily hourly concentrations at each receptor.

1.8 Why Do We Model?

In order to maintain air quality at reasonable levels many nations have air quality standards that are allowed by regulation to be exceeded only rarely. It is necessary to obtain a permit for construction of a new facility in most countries. Part of the permit application is to show that the new facility as completed and operating will not violate the air quality standards for each regulated pollutant.

Since it is not possible to make measurements of resulting air quality for a facility that has not yet been constructed, air quality dispersion modeling is about the only way to estimate this future impact.

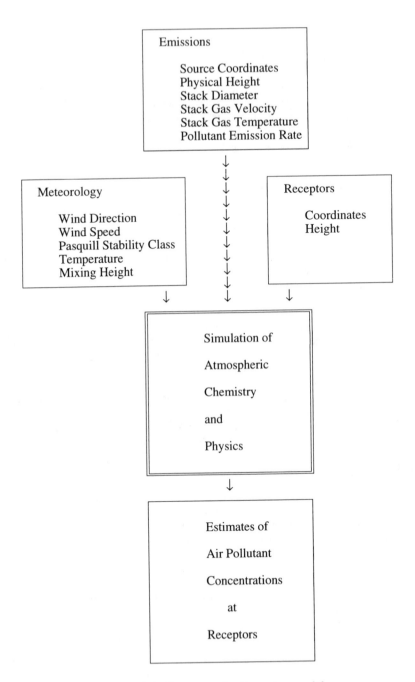

Figure 1.4 Structure of a dispersion model.

CHAPTER 2.

ESTIMATES OF ATMOSPHERIC DISPERSION

In this chapter the basic methods used in making air pollutant concentration estimates from sources releasing continuously, using the Gaussian dispersion equations, will be discussed. It is assumed that the source and the receptor positions at which concentration estimates are to be made are on or above a level surface, that is, flat terrain.

2.1 Assumptions in Gaussian Modeling

In order to estimate pollutant concentrations using the equations presented in this chapter, several assumptions are made.

- Continuous Emissions

The emissions of pollutant in mass per time are taking place continuously and the rate of these emissions are not variable over time.

- Conservation of Mass

During the transport of pollutants from source to receptor, the mass that is emitted from the source is assumed to remain in the atmosphere. None of the material is removed through chemical reaction nor is lost at the ground surface through reaction, gravitational settling, or turbulent impaction. It is assumed that any of the released pollutant that is dispersed close to the ground surface by turbulent eddies is again dispersed away from the ground surface by other subsequent turbulent eddies. This is called eddy reflection.

- Steady-State Conditions

The meteorological conditions are assumed to persist unchanged with time, at least over the time period of transport (travel time) from source to receptor. It is very easy to satisfy this assumption for close in receptors under usual conditions. However, for light wind conditions or receptors at great distances, this assumption may not be satisfied.

- Crosswind and Vertical Concentration Distributions

It is assumed that the time averaged (over about one hour) concentration profiles at any distance in the crosswind direction, horizontal (perpendicular to the path of transport) are well represented by a Gaussian, or normal, distribution and, similarly, concentration profiles in the vertical direction (also perpendicular to the path of transport) are also well represented by a Gaussian, or normal, distribution.

2.2 The Gaussian Distribution

The Gaussian or normal distribution, familiar in statistics, is used to describe the crosswind and vertical distributions that result from the turbulent mixing that causes dispersion. The height of this bell-shaped curve is described by the following function:

$$f(x) = \frac{1}{\sigma \, (2 \, \pi)^{1/2}} \; \exp \left[- \frac{(x - \mu)^2}{2 \, \sigma^2} \right]$$

where μ is the x position where the center of the distribution occurs, that is, the x corresponding to the location of the peak of $f(x)$. The mathematical notation exp(a) is used, which is a notation for the mathematical constant e, the base of the natural logarithms, which has a value of 2.71828... raised to the power a. This was probably initiated by some professor's secretary whose typewriter roller was not capable of half line spacing, thus making the typing of exponents difficult. The magnitude of the peak of the distribution is 0.398942.. or $1/(2 \, \pi)^{1/2}$. The area under the curve integrates to 1. The shape of the distribution in the horizontal, whether narrow or broad, is determined by the magnitude of the standard deviation, σ. It may be of interest to note that the above function is part of the tribute to Carl Friedr. Gauss (1777-1855) that appears on the 10 German Mark Banknote issued by the Deutsch Bundesbank on 2 January 1989.

The distribution is shown as Figure 2.1. Values of the exponential portion of the above function, $f(x)$, which varies from 0 to 1, are given in Table 2.1 (at the end of the chapter), where s is the ratio of x/σ, and μ is zero. Other statistical distributions that fall off from a central value with distance might have been used. The principal disadvantage of the Gaussian distribution is that it extends from $-\infty$ on one side to $+\infty$ on the other. Real plume spreading will be finite. However, from a practical standpoint, as can be seen from Figure 2.1 and from Table 2.1, the height of the Gaussian distribution is very small beyond the limits of $\pm 4 \, \sigma$. Therefore, the use of the Gaussian distribution is quite satisfactory.

2.3 Coordinate System

A right-hand coordinate system (Figure 2.2) can be specified with three orthogonal axes, x, y, and z, with the origin of the system at groundlevel at the point of emission (for groundlevel releases) or directly beneath the point of emission (for elevated releases). It is assumed that the x axis is oriented in the direction of transport of the centerline of the pollutant plume, that is, on the ground directly beneath the centerline of the plume. The y axis is crosswind and the z axis vertical. This orientation of the axis system is assumed to exist for the time period equivalent to that of the averaging of the wind direction. This would normally be for one hour.

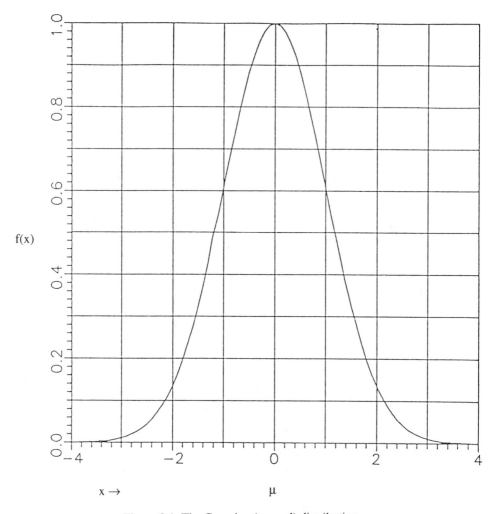

Figure 2.1 The Gaussian (normal) distribution.

2.4 Variables and Basic Dispersion Equation

The most general form of the Gaussian dispersion equation will be presented first, followed by the derivation of equations for more specialized conditions. This equation estimates the concentration at the receptor located at x downwind, y crosswind, and at a height z above the ground that results from an emission that has an effective height H above the ground.

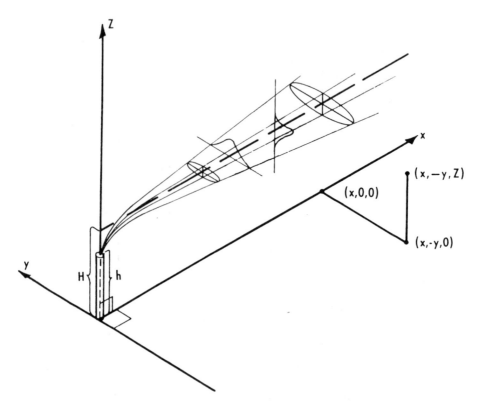

Figure 2.2   Coordinate system showing Gaussian distributions in the horizontal and vertical.

The variables used are:

χ	Air pollutant concentration in mass per volume, usually $g\ m^{-3}$.
Q	Pollutant emission rate in mass per time, usually $g\ s^{-1}$.
u	Wind speed at the point of release, $m\ s^{-1}$.
σ_y	The standard deviation of the concentration distribution in the crosswind direction, m, at the downwind distance x.
σ_z	The standard deviation of the concentration distribution in the vertical direction, m, at the downwind distance x.
π	The mathematical constant pi equal to 3.1415926... .
H	The effective height of the centerline of the pollutant plume.

The notation used following χ in parentheses is to give the three coordinates of the receptor location according to the coordinate scheme described above. Following a semicolon, the effective height of emission of the source is given.

The equation is given as four separate factors which are multiplied times each other. These four factors represent the dependency upon emissions, or the source factor, and what occurs in the three dimensions parallel to the three coordinate axes.

$$\chi(x,y,z;H) =$$

Emissions factor $\qquad Q$

Downwind factor $\qquad \dfrac{1}{u}$

Crosswind factor

$$\frac{1}{(2\pi)^{1/2}\,\sigma_y}\;\exp\left[-\frac{y^2}{2\,\sigma_y{}^2}\right]$$

Vertical factor

$$\frac{1}{(2\pi)^{1/2}\,\sigma_z}\left\{\exp\left[-\frac{(H-z)^2}{2\,\sigma_z{}^2}\right] + \exp\left[-\frac{(H+z)^2}{2\,\sigma_z{}^2}\right]\right\}$$

$$(2.1)$$

A brief explanation of the four terms follows.

1. The concentrations at the receptor are directly proportional to the emissions.

2. Parallel to the x axis, the concentrations are inversely proportional to wind speed as explained in Chapter 1.

3. Parallel to the y axis, that is, crosswind, the concentrations are inversely proportional to the crosswind spreading, σ_y, of the plume; the greater the downwind distance from the source, the greater the horizontal spreading, σ_y, the lower the concentration. The exponential involving the ratio of y to σ_y just corrects for how far off the center of the distribution the receptor is in terms of standard deviations. The receptor is y from the center since the crosswind distribution center is at $y = 0$, that is, directly above the x-axis.

4. Parallel to the z axis, that is, vertical, the concentrations are inversely proportional to the vertical spreading of the plume, σ_z; the greater the downwind distance from the source, the greater the vertical dispersion and the lower the concentration. The sum of the two exponential terms in the vertical factor represent how far the receptor height, z, is

from the plume centerline in the vertical. The first term represents the direct distance, H - z, of the receptor from the plume centerline. The second term represents the eddy reflected distance of the receptor from the plume centerline, which is the distance from the centerline to the ground, H, plus the distance back up to the receptor, z, after eddy reflection.

After doing the multiplication the equation simplifies to:

$$\chi(x,y,z;H) = \frac{Q}{2\pi u\,\sigma_y\,\sigma_z}\,\exp\left[-\frac{y^2}{2\sigma_y^2}\right]$$
$$\left\{\exp\left[-\frac{(H-z)^2}{2\sigma_z^2}\right] + \exp\left[-\frac{(H+z)^2}{2\sigma_z^2}\right]\right\} \tag{2.1}$$

2.5 Derivation of Additional Equations for Specific Situations

For receptors at groundlevel, z = 0, the above equation reduces to:

$$\chi(x,y,0;H) = \frac{Q}{\pi u\,\sigma_y\,\sigma_z}\,\exp\left[-\frac{y^2}{2\sigma_y^2}\right]\exp\left[-\frac{H^2}{2\sigma_z^2}\right] \tag{2.2}$$

In order to make concentration estimates directly beneath the plume centerline, y = 0, at groundlevel, z = 0, the equation further reduces to:

$$\chi(x,0,0;H) = \frac{Q}{\pi u\,\sigma_y\,\sigma_z}\,\exp\left[-\frac{H^2}{2\sigma_z^2}\right] \tag{2.3}$$

To calculate concentrations at the plume centerline, y = 0, z = H, equation 2.1 becomes:

$$\chi(x,0,H;H) = \frac{Q}{2\pi u\,\sigma_y\,\sigma_z}\,\left\{1 + \exp\left[-\frac{(2H)^2}{2\sigma_z^2}\right]\right\} \tag{2.4}$$

To calculate concentrations along the plume centerline at groundlevel from a groundlevel release, y = 0, z = 0, H = 0, eq. 2.1 becomes:

$$\chi(x,0,0;0) = \frac{Q}{\pi u\,\sigma_y\,\sigma_z} \tag{2.5}$$

See Chapter 5 relative to the use of the Gaussian equation for groundlevel sources.

2.6 Pasquill Stability Class

Pasquill (1961) introduced a method of estimating the atmospheric stability, incorporating considerations of both mechanical and buoyant turbulence. The major features of this method are given in Table 2.2. The mechanical turbulence is considered by the inclusion of the surface (approximately 10-meter above ground) wind speed. The positive generation of buoyant turbulence is considered through the insolation (incoming solar radiation). The negative generation of buoyant turbulence is considered through the nighttime cloud cover. The less the cloud cover the greater the amount of heat that escapes from the surface through infrared radiation. High wind speeds or overcast cloudiness will produce neutral conditions, D class stability. Unstable conditions are strongly unstable, A; moderately unstable, B; and slightly unstable, C. Stable conditions are slightly stable, E; and moderately stable, F. The hyphens at low wind speeds at night can be considered strongly stable, and sometimes are referred to as "G".

Table 2.2 Key to Pasquill stability categories.
Source: From Pasquill, 1961.

Surface wind speed (at 10 m) m s^{-1}	Insolation			Night	
	Strong	Moderate	Slight	Thinly overcast or > 4/8 low cloud	< 3/8 cloud
< 2	A	A - B	B	—	—
2 - 3	A - B	B	C	E	F
3 - 5	B	B - C	C	D	E
5 - 6	C	C - D	D	D	D
> 6	C	D	D	D	D

Notes:

1. Strong Insolation corresponds to sunny midday in midsummer in England; slight insolation to similar conditions in midwinter.

2. Night refers to the period from 1 hour before sunset to 1 hour after sunrise.

3. The neutral category D should also be used, regardless of wind speed, for overcast conditions during day or night and for any sky conditions during the hour preceding or following night as defined above.

2.7 Pasquill-Gifford Dispersion Parameters

Pasquill (1961) suggested that to estimate dispersion one should measure the horizontal and vertical fluctuations of the wind. If there are no measurements of wind fluctuations, another method is presented in Pasquill (1961). The horizontal angular spreading of the plume at two distances downwind from the source, for different stabilities, and a graphical presentation of the height of the plume, also at various distances downwind for different stabilities, are given along with equations to determine downwind concentrations. It is stated that the technique is assuming horizontal and vertical concentration distributions that are Gaussian. Gifford (1960a) converted the horizontal plume widths to σ_y and the plume heights to σ_z. Plots of these dispersion parameters on a logarithmic scale as a function of downwind distance from source to receptor, also on a logarithmic scale, are given in Figures 2.3 and 2.4. These values are considered applicable for rural conditions and are commonly referred to as the Pasquill-Gifford parameters. Note that these are considered to be functions of only downwind distance and Pasquill stability class. Tables 2.3 and 2.4 provide the equations that are used to determine the values. Table 2.5 (at the end of the chapter) gives values for a number of downwind distances.

Table 2.3 Pasquill-Gifford horizontal dispersion parameters.

$$\sigma_y = 1000 \ x \ \tan(T)/2.15$$

where x is downwind distance in km and T is one-half Pasquill's θ in degrees. T, as a function of x, is determined for each stability from the following:

Stability	Equation for T
A	$T = 24.167 - 2.5334 \ \ln(x)$
B	$T = 18.333 - 1.8096 \ \ln(x)$
C	$T = 12.5 - 1.0857 \ \ln(x)$
D	$T = 8.3333 - 0.72382 \ \ln(x)$
E	$T = 6.25 - 0.54287 \ \ln(x)$
F	$T = 4.1667 - 0.36191 \ \ln(x)$

2.8 Averaging Time of the Pasquill-Gifford Dispersion Parameters

If one measures the wind direction over a brief time interval of one or two minutes, the range of directions will be rather limited resulting in the directions of plume transport to be limited also. For higher wind speeds, such as greater than 10 m s^{-1}, the wind directions will continue to be rather limited as the time period of measurement increases. However, for lighter wind speeds, greater excursions of wind direction from the mean will be sampled as the averaging period increases. From continuous sources this increased variation will cause downwind moving plumes to occur throughout a sector of

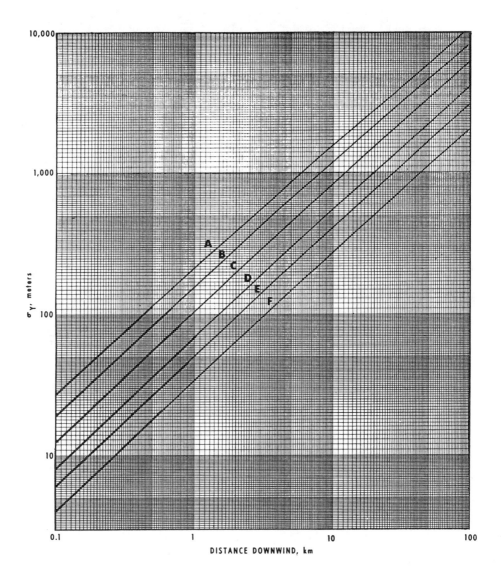

Figure 2.3 Pasquill-Gifford horizontal dispersion parameter, σ_y, as functions of Pasquill stability class and downwind distance from the source.

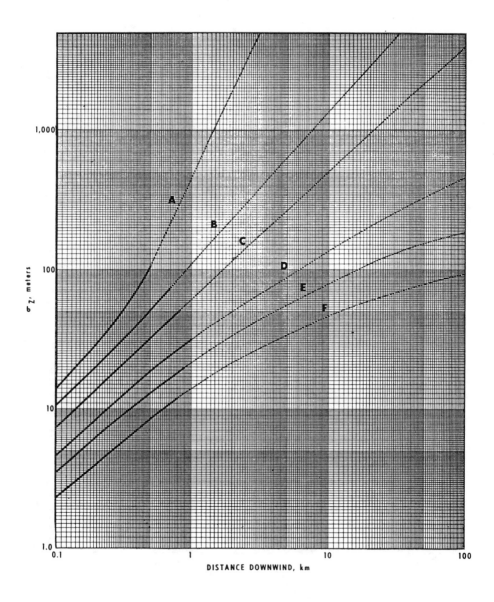

Figure 2.4   Pasquill-Gifford vertical dispersion parameter, σ_z, as functions of Pasquill stability class and downwind distance from the source.

greater angular extent. The result is an increased horizontal dispersion of the time-averaged plume as the sampling period increases.

In Pasquill (1961), the paper that defines the parameters that have become known as the Pasquill-Gifford parameters, Pasquill indicates on p. 39 that these parameters are for an averaging time of a few minutes. Turner (1967) interpreted a few minutes to be about 10 minutes. Pasquill (1976) clarified that the averaging time for the σ_y was three minutes.

Table 2.4 Pasquill-Gifford vertical dispersion parameter; $\sigma_z = a\, x^b$ where x is in km.

Stability	Distance (km)			a	b	σ_z at upper boundary
A	>3.11					5000.
	0.5	–	3.11	453.85	2.1166	
	0.4	–	0.5	346.75	1.7283	104.7
	0.3	–	0.4	258.89	1.4094	71.2
	0.25	–	0.3	217.41	1.2644	47.4
	0.2	–	0.25	179.52	1.1262	37.7
	0.15	–	0.2	170.22	1.0932	29.3
	0.1	–	0.15	158.08	1.0542	21.4
	<0.1			122.8	0.9447	14.0
B	>35.					5000.
	0.4	–	35.	109.30	1.0971	
	0.2	–	0.4	98.483	0.98332	40.0
	<0.2			90.673	0.93198	20.2
C	all x			61.141	0.91465	
D	>30.			44.053	0.51179	
	10.	–	30.	36.650	0.56589	251.2
	3.	–	10.	33.504	0.60486	134.9
	1.	–	3.	32.093	0.64403	65.1
	0.3	–	1.	32.093	0.81066	32.1
	<0.3			34.459	0.86974	12.1
E	>40.			47.618	0.29592	
	20.	–	40.	35.420	0.37615	141.9
	10.	–	20.	26.970	0.46713	109.3
	4.	–	10.	24.703	0.50527	79.1
	2.	–	4.	22.534	0.57154	49.8
	1.	–	2.	21.628	0.63077	33.5
	0.3	–	1.	21.628	0.75660	21.6
	0.1	–	0.3	23.331	0.81956	8.7
	<0.1			24.260	0.83660	3.5

Table 2.4 (cont.) Pasquill-Gifford vertical dispersion parameter.

Stability	Distance (km)			a	b	σ_z at upper boundary
F	>60.			34.219	0.21716	
	30.	–	60.	27.074	0.27436	83.3
	15.	–	30.	22.651	0.32681	68.8
	7.	–	15.	17.836	0.4150	54.9
	3.	–	7.	16.187	0.4649	40.0
	2.	–	3.	14.823	0.54503	27.0
	1.	–	2.	13.953	0.63227	21.6
	0.7	–	1.	13.953	0.68465	14.0
	0.2	–	0.7	14.457	0.78407	10.9
	<0.2			15.209	0.81558	4.1

Due to wind direction meander, one would expect that frequently the σ_y for longer averaging times would be larger than those appropriate for three minutes. There is little long-period meander in the vertical (although under strong convective conditions conducive to the formation of thermals the updrafts and downdrafts each persist over time periods in excess of five minutes). However, for most conditions there is almost no long-period meander in the vertical. Pasquill (1976) indicates that the sampling time for σ_z is in excess of 10 minutes.

In the models that are approved for regulatory use by the modeling guidelines (EPA, 1986; EPA, 1987b) the Pasquill-Gifford parameters are used directly for making concentration estimates for one-hour periods for sources in rural areas. This is appropriate for periods with steady winds over one-hour periods. For example, the calculated concentrations are appropriate for periods with 20 three-minute periods that have the same mean wind direction. Since many of the models approved for regulatory use are used to estimate extreme concentrations, the use of the Pasquill-Gifford parameters will assist in estimating the higher one-hour concentrations. The exact effect of the P-G σ_y upon three-hour and 24-hour concentration estimates is largely unknown as these depend upon both the hourly horizontal dispersion and the variation in wind direction from hour to hour.

2.9 Determining Groundlevel Concentration with Downwind Distance

If the dispersion parameters were simple functions with distance, the distance of the maximum could be determined by differentiating eq 2.3 with respect to x and setting the result equal to zero. However, as seen from Figures 2.3 and 2.4, the dispersion parameters, σ_y and σ_z are not simple functions of downwind distance. Therefore the determination of the distance to the maximum and the maximum concentration is accomplished by trial and error by evaluating eq 2.3 for numerous downwind distances.

A systematic step-wise procedure is followed in the program on the floppy diskette (see Chapter 7) to determine the maximum concentration and the distance of its occurrence.

2.10 Nomogram for Determination of Distance to Groundlevel Maximum and Maximum Concentration

By repeated systematic solutions to eq. 2.3 at a number of distances for various pairs of Pasquill stability class and effective heights of release using the Pasquill-Gifford dispersion parameters, pairs of maximum relative concentration (relative to the emissions) normalized for wind speed, $\chi u/Q$, and distance to maximum concentration have been determined. These results have been summarized in the nomogram (Figure 2.5). The horizontal scale of the nomogram is a log scale of the distance to maximum concentration. The vertical scale of maximum $\chi u/Q$ is also a log scale. The lines running from upper left to lower right are for each stability class. The connected line segments that run more or less horizontal are for different effective heights of release, H. To determine the distance to the maximum and the maximum concentration for a given effective height of release and Pasquill stability class, it is necessary to know the emission rate, Q, and the horizontal wind speed past the point of release. (It is assumed that the effective height of release, H, is the proper height to determine this wind speed.) Find the intersection of the lines corresponding to the stability class and the effective height of release. If the effective height does not correspond exactly to a line with a given value, it will be necessary to interpolate for the proper position along the stability line. Note that the spacing of the effective height of release lines is approximately logarithmic. When the proper point is found, moving vertically down to the x-axis will determine the distance to the maximum concentration in kilometers. By moving horizontally to the left the maximum $\chi u/Q$ is determined. The maximum concentration is determined by multiplying the value of $\chi u/Q$ found on the vertical scale by Q/u.

For example, for a source release rate of 10 g s^{-1} from an effective height of 20 meters at a wind speed of 4 m s^{-1}, and a Pasquill stability class B, the distance to the maximum concentration is read as 0.14 km, on the horizontal axis directly beneath the intersecting point of the B stability line and the 20-meter line segments. The maximum $\chi u/Q$ is 3.1 x 10^{-4} read from the vertical axis horizontally to the left of the previously described point. The maximum concentration is found by multiplying this value by Q/u, 10/4 = 2.5. The resulting maximum concentration is 7.8 x 10^{-4} g m^{-3} or 780 μg m^{-3}.

2.11 Accuracy of Estimates

The estimation of the magnitude of the concentration, χ, that occurs at some point downwind for a given x is considered to be a "best estimate." The errors in the emission rate, Q, will propagate directly into an error in the calculated concentration. Since wind speed generally increases with height above the ground, the estimation of the wind speed at the point of release may be in error on the order of 10 to 15 percent. Since the effective height of release, H, is dependent upon wind speed and stack parameters, it may also have errors of up to 20 to 25 percent. With regard to the dispersion parameters, it must be kept in mind that by using the Pasquill stabilities and the Pasquill-Gifford

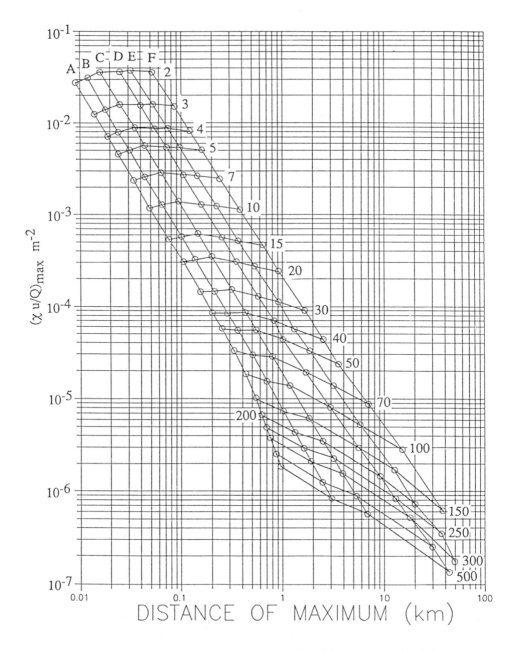

Figure 2.5   Relative maximum concentration normalized for wind speed and distance to maximum concentration as functions of Pasquill stability class and effective height of release.

dispersion parameters we are considering the atmosphere in only six classes while, in reality, it is a continuum. One will note by making sample calculations that considerably different concentrations are calculated with a change of one stability class in the assumptions. The difference is greater at greater distances.

Larger errors in the dispersion parameters, σ_y and σ_z, are expected for the extremes of stability and at larger distances. In some cases the σ_z may be expected to be correct within a factor of two. These are: 1) all stabilities for distance of travel out to a few hundred meters; 2) neutral to moderately unstable conditions for distances out to a few kilometers; and 3) unstable conditions in the lower 1000 meters of the atmosphere, with an inversion limiting the mixing above, for distances out to 10 km or more. Uncertainties in the estimates of σ_y are, in general, less than those of σ_z. The groundlevel centerline concentrations for these three cases (where σ_z can be expected to be within a factor of 2) should be correct within a factor of 3, including errors in σ_y, H, and u. The relative confidence in the σ's is indicated by the solid and dashed lines in Figures 2.3 and 2.4.

It should be noted that the σ_y behavior may not be as neat and orderly as given in Figure 2.3 especially during periods of light winds. There may be some meander (wind direction changes with a longer time period) of the wind under such conditions which will cause an increase in the effective horizontal dispersion that is not really due to turbulent fluctuations. This meander under light wind conditions is not included in the Pasquill-Gifford σ_y's.

The errors that have been discussed are those associated with estimates of the concentrations directly downwind from the point of release. Slight errors in the estimation of wind direction, especially under stable conditions when pollutant plumes are relatively narrow, can result in tremendous errors of concentration where the problem is to estimate the concentration at specific locations. This is also the principal reason why so many hour-to-hour field concentration measurements relate rather poorly with concentration estimates. The estimated plume path dependent on the estimated wind direction is somewhat different than the actual plume path responding to the actual wind direction at the height of the plume centerline. In these cases the magnitude of the highest downwind concentrations under the stated stability and wind speed are estimated quite well, but the location of this maximum may be in error. Therefore, if one is trying to use dispersion estimates to estimate the concentration at specific times and specific locations, it is important to try to make exceptionally good estimates of the wind direction for each time period or expect to put up with large error bounds, perhaps as much as a factor of ten, about the estimated concentrations. See Problem 5e in Chapter 8.

2.12 Determining the Distance to a Concentration Level of Concern for a Groundlevel Release

For a groundlevel release the equation for concentration directly downwind (y = 0) is given by eq. 2.5. In order to determine the distance to a particular value of concentration, call it the Level-of-Concern concentration, χ_{LOC}, this equation can be rearranged to solve for the product of σ_y times σ_z. This is:

$$\sigma_y \, \sigma_z = \frac{Q}{\pi \, u \, \chi_{LOC}} \tag{2.6}$$

The distance where the product achieves this value can then be approximated by inspection of the right side of Table 2.5 or of Figure 2.6. Of course, the emission rate, Q, and wind speed, u, must be known or closely approximated.

This just gives the distance to the point where the concentration can be expected to drop off to the level of concern. As stated in the above section, the location where this is occurring is highly dependent upon the wind direction. If nothing is known about the wind direction, all that can be said is that there is a circle with a radius equal to the distance to the level-of-concern concentration and the concentrations that are higher are occurring somewhere within this circle. If something more definitive is known about the wind direction at this site for this time interval then the location of the high concentrations can be located more specifically.

The distance to χ_{LOC} can be determined directly from the above procedures for the simplified situation of the groundlevel release. For an elevated release the additional complication of the exponential involving the ratio of H to σ_z occurs. A direct solution is not available and eq. 2.3 must be solved at various downwind distances to determine where the concentration decreases to the LOC. The first estimate can be made with eq. 2.6 however, and then closer distances tried until the proper distance is found.

2.13   Treatment of Effect of Mixing Height

The mixing height according to Holzworth (1972, p 3) is "the height above the surface through which relatively vigorous mixing occurs." Therefore the mixing height is assumed to occur with unstable and neutral conditions and to be undefined when the surface layer is stable. Plume trapping occurs when the plume is trapped between the ground surface and a stable layer aloft. Such a stable layer frequently caps the mixing height. Bierly and Hewson (1962) have suggested the use of an equation that accounts for the multiple eddy reflections from both the ground and the stable layer where z_i is the height of the stable layer and J = 3 or 4 is sufficient to include the reflections of any significance. The principal off-axis position vertically involving the mixing height is the distance from the point of release to the mixing height plus the distance from the mixing height to the receptor height. However, all other possible combinations of multiple eddy reflection between the ground and the mixing height have been included in eq. 2.7. This equation is evaluated for receptors that are close to the source.

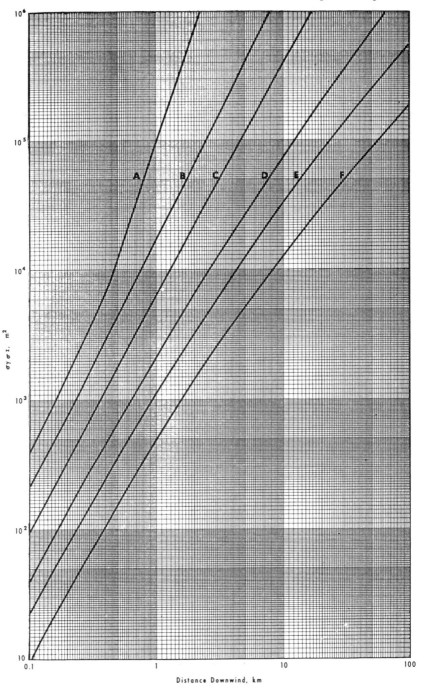

Figure 2.6 The product of $\sigma_y\sigma_z$ as a function of downwind distance from the source.

$$\chi(x,0,z;H) = \frac{Q}{2 \pi u \, \sigma_y \, \sigma_z} \left(\exp\left[-\frac{(H-z)^2}{2 \, \sigma_z^2} \right] + \exp\left[-\frac{(H-z)^2}{2 \, \sigma_z^2} \right] \right.$$

$$+ \sum_{N=1}^{N=J} \left\{ \exp\left[-\frac{(z-H-2Nz_i)^2}{2 \, \sigma_z^2} \right] + \exp\left[-\frac{(z+H-2Nz_i)^2}{2 \, \sigma_z^2} \right] \right.$$

$$\left. + \exp\left[-\frac{(z-H+2Nz_i)^2}{2 \, \sigma_z^2} \right] + \exp\left[-\frac{(z+H+2Nz_i)^2}{2 \, \sigma_z^2} \right] \right\} \left. \vphantom{\sum} \right) \qquad (2.7)$$

For receptors that are quite some distance from the source, the vertical term becomes just $1/z_i$, a uniform mixing between the ground and the mixing height as shown in eq. 2.8. It has been found through calculations that regardless of effective height of emission or receptor height, provided that both are between the ground and the mixing height, if the value of σ_z exceeds 1.6 times the mixing height then the solutions of eq. 2.7 will yield a uniform concentration with height between the ground and the mixing height and the following simplified equation can be used.

$$\chi(x,0,z;H) = \frac{Q}{(2\pi)^{1/2} \, u \, \sigma_y \, z_i} \qquad (2.8)$$

A good approximation to the use of equation 2.7 can be made by assuming no effect of the stable layer until $\sigma_z = 0.47 \, (z_i - H)$. It is assumed that at this distance, x_i, the stable layer begins to affect the vertical distribution so that at the downwind distance $2 \, x_i$, uniform vertical mixing has taken place and equation 2.8 can be used. For distance between x_i and $2 \, x_i$ a good approximation to the groundlevel centerline concentration is that read from a straight line drawn between the concentrations for distances x_i and $2 \, x_i$ on a log-log plot of groundlevel centerline concentration as a function of downwind distance.

Holzworth (1972) analyzed radiosonde data to estimate mixing heights throughout the year. Figure 2.7 gives the mean winter afternoon mixing height for the 48 contiguous states. Figure 2.8 gives the mean summer afternoon mixing height.

2.14   Graphs for Estimates of Downwind Groundlevel Concentrations

To avoid repetitive computations, Figures 2.9A through F give relative groundlevel concentrations times wind speed, $\chi u/Q$, as functions of downwind distance, x, for various effective heights of emission and mixing heights for each stability class. Estimates of concentration may be determined by multiplying ordinate values by Q/u.

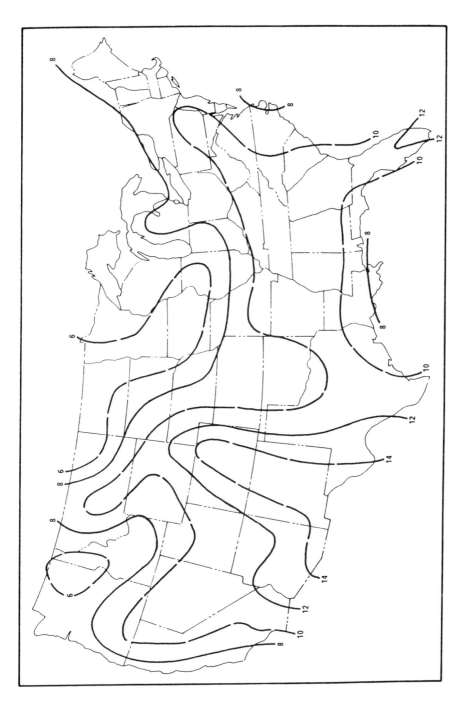

Figure 2.7 Mean winter afternoon mixing heights in hundreds of meters. Source: Holzworth(1972).

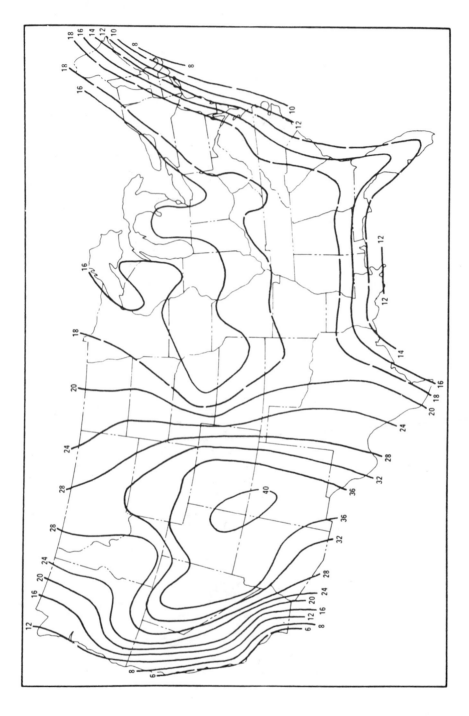

Figure 2.8 Mean summer afternoon mixing heights in hundreds of meters. Source: Holzworth(1972).

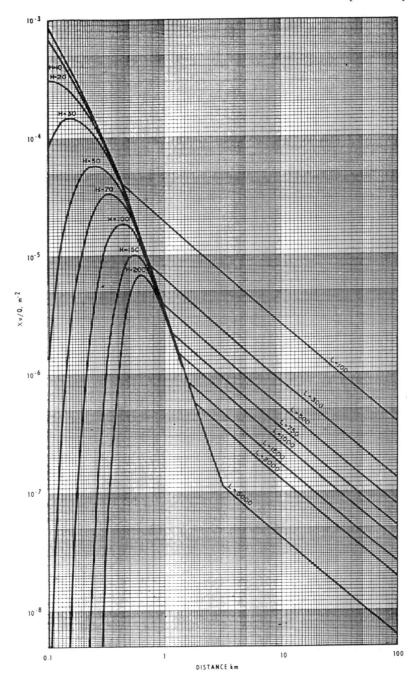

Figure 2.9A   χu/Q with distance for various heights of emission (H) and limits to vertical dispersion, A stability.

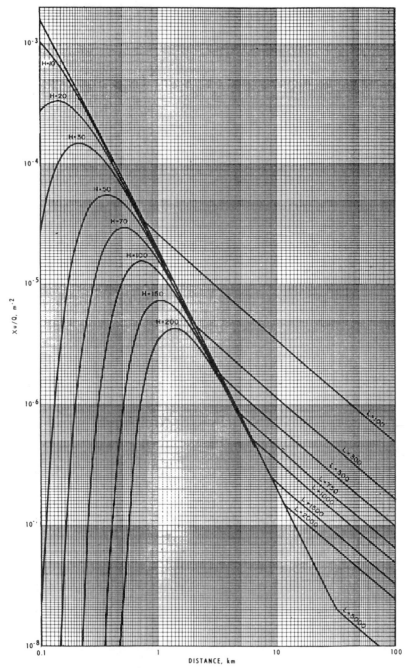

Figure 2.9B   χu/Q with distance for various heights of emission (H) and limits to vertical dispersion, B stability.

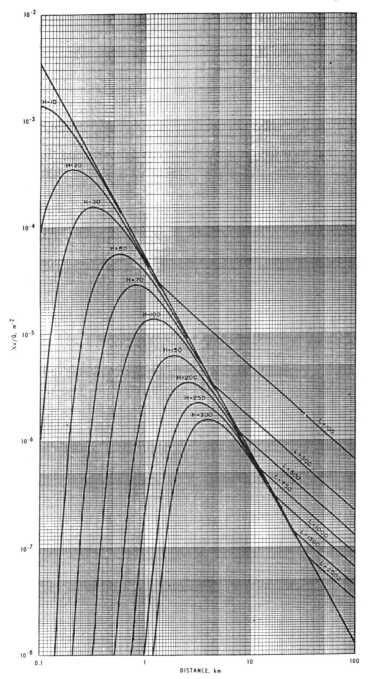

Figure 2.9C   χu/Q with distance for various heights of emission (H) and limits to vertical dispersion, C stability.

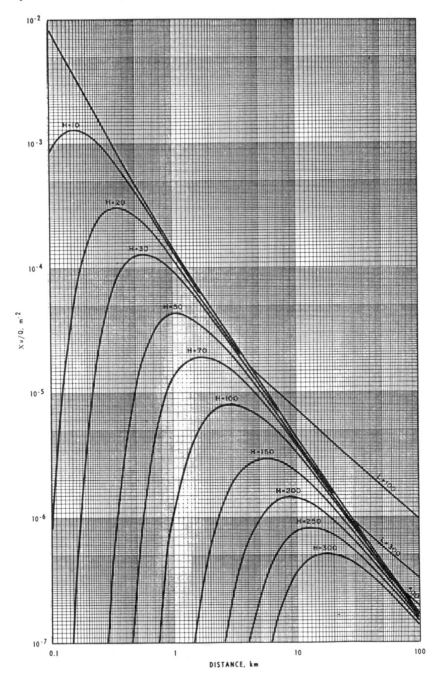

Figure 2.9D   χu/Q with distance for various heights of emission (H) and limits to vertical dispersion, D stability.

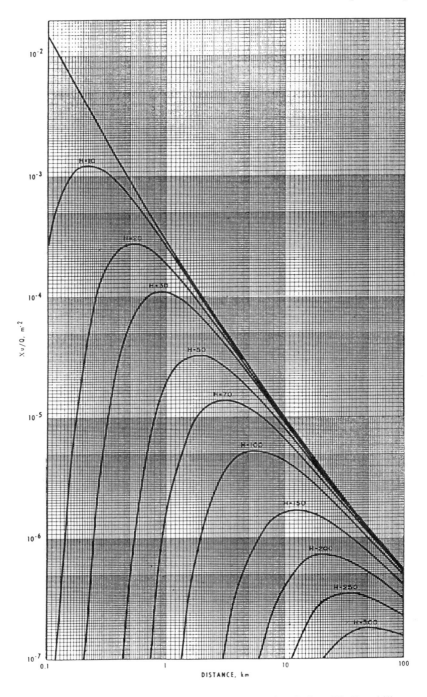

Figure 2.9E   χu/Q with distance for various heights of emission (H), E stability.

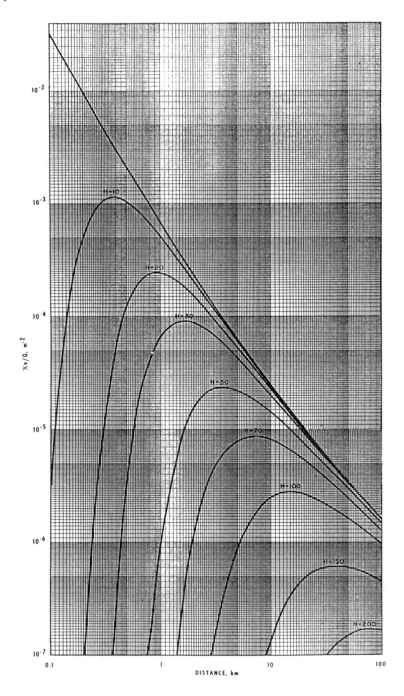

Figure 2.9F   χu/Q with distance for various heights of emission (H), F stability.

2.15 Determining Groundlevel Concentration Isopleths

Often one wishes to determine the locations where concentrations equal or exceed a given magnitude. First, the axial position of the plume must be determined by the mean wind direction. For plotting isopleths of groundlevel concentrations, the relationship between groundlevel centerline concentrations and groundlevel off-axis concentrations can be used:

$$\frac{\chi(x,y,0;H)}{\chi(x,0,0;H)} = \exp\left[-\frac{y^2}{2\,\sigma_y^2}\right] \tag{2.9}$$

The y coordinate of a particular isopleth from the x-axis can be determined at each downwind distance x. Suppose that one wishes to know the off-axis distance to the 10^{-3} g m^{-3} isopleth at an x of 600 m, under Pasquill stability B, where the groundlevel concentration directly downwind of the source at this distance is 2.9 x 10^{-3} g m^{-3}.

$$\exp\left[-\frac{y^2}{2\,\sigma_y^2}\right] = \frac{\chi(x,y,0;H)}{\chi(x,0,0;H)} = \frac{10^{-3}}{2.9\ x\ 10^{-3}} = 0.345 \tag{2.10}$$

From Table 2.1, when exp $[-0.5\ (y/\sigma_y)^2] = 0.345$,

$$y/\sigma_y = 1.46$$

From Table 2.5, for stability B and x = 600 m, σ_y = 97.5. Therefore y = 1.46 x 97.5 = 142 m. This is the distance of the 10^{-3} isopleth from the x-axis at a downwind distance of 600 m. This can also be determined from:

$$y = \left\{2\ln\left[\frac{\chi(x,0,0;H)}{\chi(x,y,0;H)}\right]\right\}^{1/2}\sigma_y \tag{2.11}$$

The position corresponding to the downwind distance and off-axis distance can then be plotted. After a number of points have been plotted for different downwind distances, the concentration isopleth may be drawn (see problems 5c and 8 in Chapter 8). Figures 2.10 A - F gives groundlevel isopleths of $\chi u/Q$ for various stabilities for releases at groundlevel (H = 0), and Figure 2.11 A - F gives groundlevel isopleths for an effective height of release of 100 meters. For example, to locate the 10^{-3} g m^{-3} isopleth resulting from a groundlevel source of 20 g s^{-1} under B stability conditions with wind speed 2 m s^{-1}, one must first determine the corresponding value of $\chi u/Q$ since this is the quantity graphed in Figure 2.10 B. $\chi u/Q = 10^{-3}$ times 2/20 = 10^{-4}. Therefore the $\chi u/Q$ isopleth in Fig. 2-5 having a value of 10^{-4} m^{-2} corresponds to a χ isopleth with a value of 10^{-3} g m^{-3}.

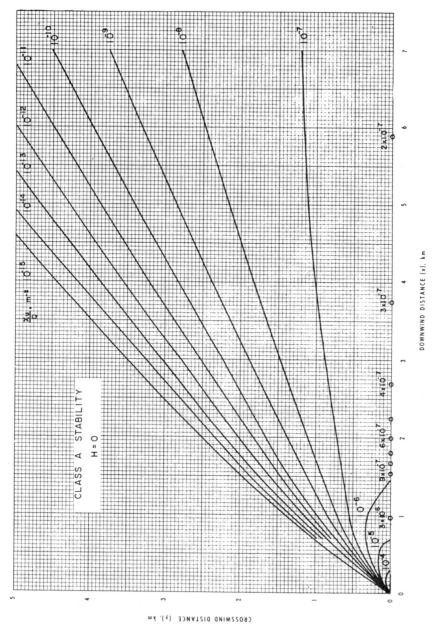

Figure 2.10A Isopleths of $\chi u/Q$ for a ground-level source, A stability.

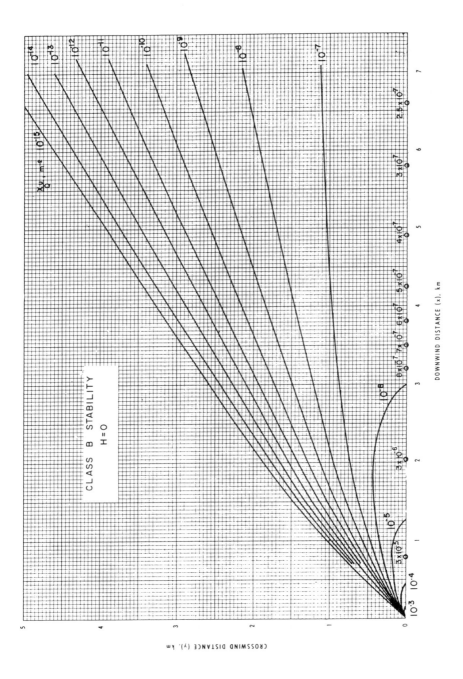

Figure 2.10B Isopleths of χu/Q for a ground-level source, B stability.

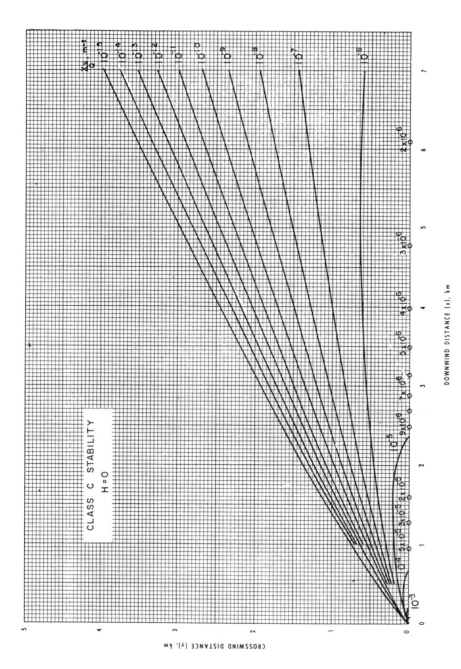

Figure 2.10C Isopleths of χu/Q for a ground-level source, C stability.

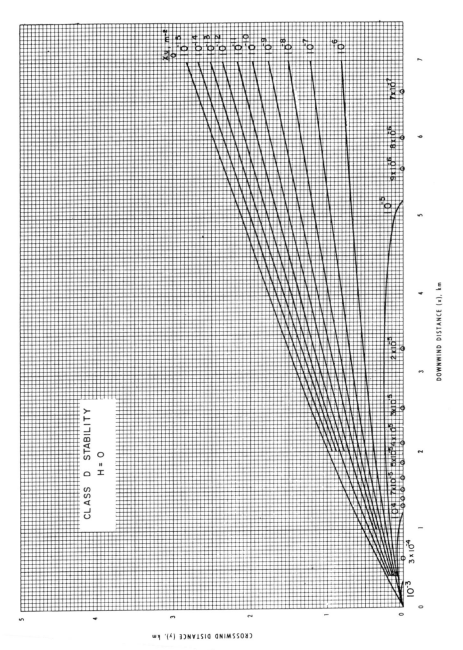

Figure 2.10D Isopleths of $\chi u/Q$ for a ground-level source, D stability.

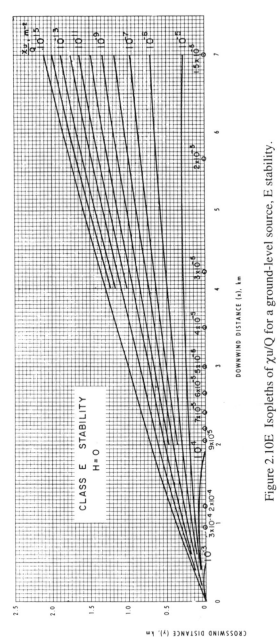

Figure 2.10E Isopleths of $\chi u/Q$ for a ground-level source, E stability.

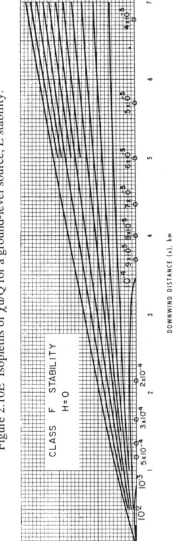

Figure 2.10F Isopleths of $\chi u/Q$ for a ground-level source, F stability.

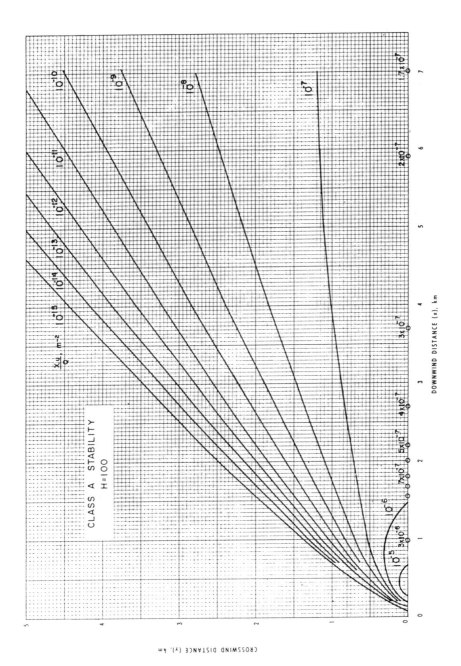

Figure 2.11A Isopleths of $\chi u/Q$ for a source 100 meters high, A stability.

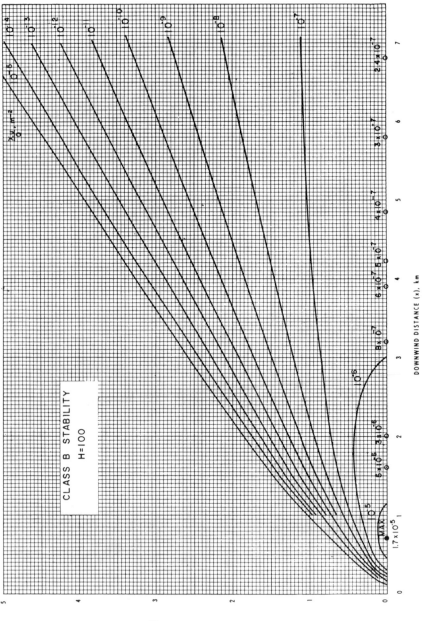

Figure 2.11B Isopleths of $\chi u/Q$ for a source 100 meters high, B stability.

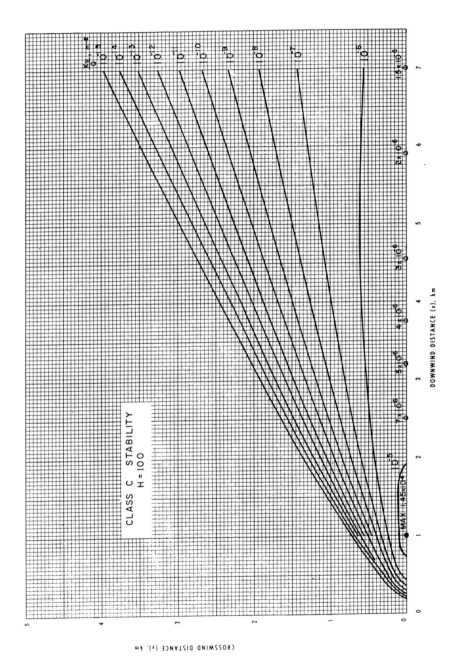

Figure 2.11C Isopleths of χu/Q for a source 100 meters high, C stability.

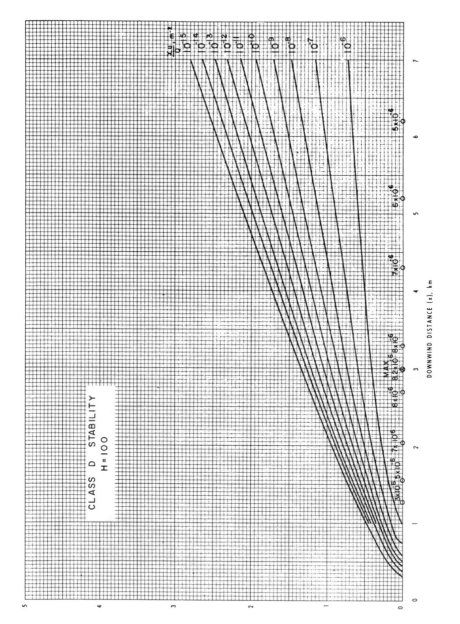

Figure 2.11D Isopleths of χu/Q for a source 100 meters high, D stability.

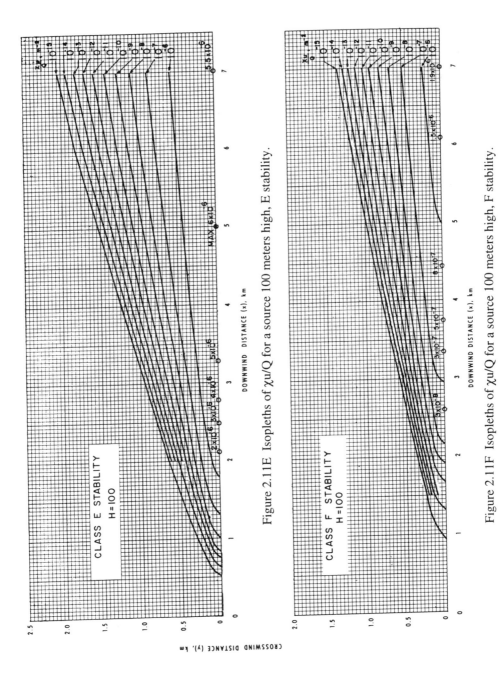

Figure 2.11E Isopleths of χu/Q for a source 100 meters high, E stability.

Figure 2.11F Isopleths of χu/Q for a source 100 meters high, F stability.

2.16   Areas Within Isopleths

Figure 2.12 gives areas within isopleths of groundlevel concentration in terms of $\chi u/Q$ for a groundlevel source for various stability categories (Gifford, 1962; Hilsmeier and Gifford, 1962). For the example just given, the area of the 10^{-3} g m^{-3} isopleth (10^{-4} m^{-2} times u/Q isopleth) is about 5 times 10^4 m^2.

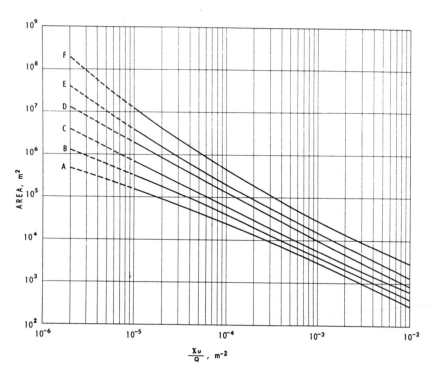

Figure 2.12 Area within isopleths for a groundlevel source
(from Hilsmeier and Gifford, 1962).

2.17   Review of Assumptions

The Gaussian dispersion that has been discussed in this chapter has been based on assumptions that need to be clearly understood:

1) The emission is continuous from the source or as a minimum the emission is taking place for a longer period than the travel time (downwind distance over wind speed) to the downwind position under consideration, so that diffusion along the upwind-downwind axis may be neglected.

2) The material being dispersed is a stable (nonreactive) gas or aerosol (less than 20 μm effective diameter) which remains suspended in the air over long periods of time.

3) The equation of continuity:

$$Q = \int_0^{+\infty} \int_{-\infty}^{+\infty} \chi \ u \ dy \ dz \tag{2.12}$$

is fulfilled, i.e., none of the material emitted is removed from the plume as it moves downwind and there is complete eddy reflection at the ground and at the mixing height.

4) The mean wind direction specifies the x-axis, and the wind speed at the height above ground of the point of release (stack top) represents the diluting wind.

5) Except where specifically mentioned, the plume constituents are distributed normally in both the crosswind and vertical directions.

6) The σ's given in Figs. 2.3 and 2.4 represent concentration averaging time periods of about three to 10 minutes.

The basics of considering dispersion by use of the Gaussian equation have been presented in this chapter.

Table 2.1 (Part 1) Solutions to Exponentials: B = exp (-0.5 Asq) where Asq is A squared, or A times A.
Note that 9.67E-01 means 9.67 times 10 to the -1 power.

A	B 0.00	0.01	0.02	0.03	0.04	0.05	0.06	0.07	0.08	0.09
0.00	1.00E+00	1.00E+00	1.00E+00	1.00E+00	9.99E-01	9.99E-01	9.98E-01	9.98E-01	9.97E-01	9.96E-01
0.10	9.95E-01	9.94E-01	9.93E-01	9.92E-01	9.90E-01	9.89E-01	9.87E-01	9.86E-01	9.84E-01	9.82E-01
0.20	9.80E-01	9.78E-01	9.76E-01	9.74E-01	9.72E-01	9.69E-01	9.67E-01	9.64E-01	9.62E-01	9.59E-01
0.30	9.56E-01	9.53E-01	9.50E-01	9.47E-01	9.44E-01	9.41E-01	9.37E-01	9.34E-01	9.30E-01	9.27E-01
0.40	9.23E-01	9.19E-01	9.16E-01	9.12E-01	9.08E-01	9.04E-01	9.00E-01	8.95E-01	8.91E-01	8.87E-01
0.50	8.82E-01	8.78E-01	8.74E-01	8.69E-01	8.64E-01	8.60E-01	8.55E-01	8.50E-01	8.45E-01	8.40E-01
0.60	8.35E-01	8.30E-01	8.25E-01	8.20E-01	8.15E-01	8.10E-01	8.04E-01	7.99E-01	7.94E-01	7.88E-01
0.70	7.83E-01	7.77E-01	7.72E-01	7.66E-01	7.60E-01	7.55E-01	7.49E-01	7.43E-01	7.38E-01	7.32E-01
0.80	7.26E-01	7.20E-01	7.14E-01	7.09E-01	7.03E-01	6.97E-01	6.91E-01	6.85E-01	6.79E-01	6.73E-01
0.90	6.67E-01	6.61E-01	6.55E-01	6.49E-01	6.43E-01	6.37E-01	6.31E-01	6.25E-01	6.19E-01	6.13E-01
1.00	6.07E-01	6.00E-01	5.94E-01	5.88E-01	5.82E-01	5.76E-01	5.70E-01	5.64E-01	5.58E-01	5.52E-01
1.10	5.46E-01	5.40E-01	5.34E-01	5.28E-01	5.22E-01	5.16E-01	5.10E-01	5.04E-01	4.98E-01	4.93E-01
1.20	4.87E-01	4.81E-01	4.75E-01	4.69E-01	4.64E-01	4.58E-01	4.52E-01	4.46E-01	4.41E-01	4.35E-01
1.30	4.30E-01	4.24E-01	4.18E-01	4.13E-01	4.07E-01	4.02E-01	3.97E-01	3.91E-01	3.86E-01	3.81E-01
1.40	3.75E-01	3.70E-01	3.65E-01	3.60E-01	3.55E-01	3.50E-01	3.44E-01	3.39E-01	3.34E-01	3.30E-01
1.50	3.25E-01	3.20E-01	3.15E-01	3.10E-01	3.06E-01	3.01E-01	2.96E-01	2.92E-01	2.87E-01	2.83E-01
1.60	2.78E-01	2.74E-01	2.69E-01	2.65E-01	2.61E-01	2.56E-01	2.52E-01	2.48E-01	2.44E-01	2.40E-01
1.70	2.36E-01	2.32E-01	2.28E-01	2.24E-01	2.20E-01	2.16E-01	2.13E-01	2.09E-01	2.05E-01	2.01E-01
1.80	1.98E-01	1.94E-01	1.91E-01	1.87E-01	1.84E-01	1.81E-01	1.77E-01	1.74E-01	1.71E-01	1.68E-01
1.90	1.64E-01	1.61E-01	1.58E-01	1.55E-01	1.52E-01	1.49E-01	1.46E-01	1.44E-01	1.41E-01	1.38E-01
2.00	1.35E-01	1.33E-01	1.30E-01	1.27E-01	1.25E-01	1.22E-01	1.20E-01	1.17E-01	1.15E-01	1.13E-01
2.10	1.10E-01	1.08E-01	1.06E-01	1.03E-01	1.01E-01	9.91E-02	9.70E-02	9.49E-02	9.29E-02	9.09E-02
2.20	8.89E-02	8.70E-02	8.51E-02	8.32E-02	8.14E-02	7.96E-02	7.78E-02	7.60E-02	7.43E-02	7.27E-02
2.30	7.10E-02	6.94E-02	6.78E-02	6.62E-02	6.47E-02	6.32E-02	6.17E-02	6.03E-02	5.89E-02	5.75E-02
2.40	5.61E-02	5.48E-02	5.35E-02	5.22E-02	5.10E-02	4.97E-02	4.85E-02	4.73E-02	4.62E-02	4.50E-02
2.50	4.39E-02	4.29E-02	4.18E-02	4.07E-02	3.97E-02	3.87E-02	3.77E-02	3.68E-02	3.59E-02	3.49E-02

Table 2.1 (Part 2) Solutions to Exponentials: B = exp (-0.5 Asq) where Asq is A squared, or A times A.
Note that 9.67E-01 means 9.67 times 10 to the -1 power.

A	0.00 B	0.01	0.02	0.03	0.04	0.05	0.06	0.07	0.08	0.09
2.60	3.40E-02	3.32E-02	3.23E-02	3.15E-02	3.07E-02	2.99E-02	2.91E-02	2.83E-02	2.76E-02	2.68E-02
2.70	2.61E-02	2.54E-02	2.47E-02	2.41E-02	2.34E-02	2.28E-02	2.22E-02	2.16E-02	2.10E-02	2.04E-02
2.80	1.98E-02	1.93E-02	1.88E-02	1.82E-02	1.77E-02	1.72E-02	1.67E-02	1.63E-02	1.58E-02	1.54E-02
2.90	1.49E-02	1.45E-02	1.41E-02	1.37E-02	1.33E-02	1.29E-02	1.25E-02	1.21E-02	1.18E-02	1.14E-02
3.00	1.11E-02	1.08E-02	1.05E-02	1.01E-02	9.84E-03	9.55E-03	9.26E-03	8.98E-03	8.71E-03	8.45E-03
3.10	8.19E-03	7.94E-03	7.69E-03	7.46E-03	7.23E-03	7.00E-03	6.79E-03	6.58E-03	6.37E-03	6.17E-03
3.20	5.98E-03	5.79E-03	5.60E-03	5.43E-03	5.25E-03	5.09E-03	4.92E-03	4.77E-03	4.61E-03	4.46E-03
3.30	4.32E-03	4.18E-03	4.04E-03	3.91E-03	3.78E-03	3.66E-03	3.54E-03	3.42E-03	3.31E-03	3.20E-03
3.40	3.09E-03	2.99E-03	2.89E-03	2.79E-03	2.69E-03	2.60E-03	2.51E-03	2.43E-03	2.35E-03	2.27E-03
3.50	2.19E-03	2.11E-03	2.04E-03	1.97E-03	1.90E-03	1.83E-03	1.77E-03	1.71E-03	1.65E-03	1.59E-03
3.60	1.53E-03	1.48E-03	1.43E-03	1.38E-03	1.33E-03	1.28E-03	1.23E-03	1.19E-03	1.15E-03	1.10E-03
3.70	1.06E-03	1.03E-03	9.89E-04	9.52E-04	9.18E-04	8.84E-04	8.51E-04	8.20E-04	7.89E-04	7.60E-04
3.80	7.32E-04	7.04E-04	6.78E-04	6.53E-04	6.28E-04	6.04E-04	5.82E-04	5.60E-04	5.38E-04	5.18E-04
3.90	4.98E-04	4.79E-04	4.61E-04	4.43E-04	4.26E-04	4.09E-04	3.93E-04	3.78E-04	3.63E-04	3.49E-04
4.00	3.35E-04	3.22E-04	3.10E-04	2.97E-04	2.86E-04	2.74E-04	2.63E-04	2.53E-04	2.43E-04	2.33E-04
4.10	2.24E-04	2.15E-04	2.06E-04	1.98E-04	1.90E-04	1.82E-04	1.75E-04	1.68E-04	1.61E-04	1.54E-04
4.20	1.48E-04	1.42E-04	1.36E-04	1.30E-04	1.25E-04	1.20E-04	1.15E-04	1.10E-04	1.05E-04	1.01E-04
4.30	9.66E-05	9.25E-05	8.86E-05	8.49E-05	8.13E-05	7.78E-05	7.45E-05	7.13E-05	6.83E-05	6.53E-05
4.40	6.25E-05	5.98E-05	5.72E-05	5.48E-05	5.24E-05	5.01E-05	4.79E-05	4.58E-05	4.38E-05	4.19E-05
4.50	4.01E-05	3.83E-05	3.66E-05	3.50E-05	3.34E-05	3.20E-05	3.05E-05	2.92E-05	2.79E-05	2.66E-05
4.60	2.54E-05	2.43E-05	2.32E-05	2.21E-05	2.11E-05	2.02E-05	1.93E-05	1.84E-05	1.75E-05	1.67E-05
4.70	1.60E-05	1.52E-05	1.45E-05	1.39E-05	1.32E-05	1.26E-05	1.20E-05	1.15E-05	1.09E-05	1.04E-05
4.80	9.93E-06	9.46E-06	9.02E-06	8.59E-06	8.19E-06	7.80E-06	7.43E-06	7.08E-06	6.74E-06	6.42E-06
4.90	6.11E-06	5.82E-06	5.54E-06	5.27E-06	5.02E-06	4.78E-06	4.55E-06	4.33E-06	4.12E-06	3.92E-06
5.00	3.73E-06	3.54E-06	3.37E-06	3.21E-06	3.05E-06	2.90E-06	2.76E-06	2.62E-06	2.49E-06	2.37E-06

Table 2.1 (Part 3) Solutions to Exponentials: B = exp (-0.5 Asq) where Asq is A squared, or A times A.
Note that 9.67E-01 means 9.67 times 10 to the -1 power.

A	B 0.00	0.01	0.02	0.03	0.04	0.05	0.06	0.07	0.08	0.09
5.10	2.25E-06	2.14E-06	2.03E-06	1.93E-06	1.83E-06	1.74E-06	1.65E-06	1.57E-06	1.49E-06	1.42E-06
5.20	1.34E-06	1.28E-06	1.21E-06	1.15E-06	1.09E-06	1.03E-06	9.82E-07	9.31E-07	8.84E-07	8.38E-07
5.30	7.95E-07	7.54E-07	7.15E-07	6.78E-07	6.42E-07	6.09E-07	5.77E-07	5.47E-07	5.18E-07	4.91E-07
5.40	4.65E-07	4.41E-07	4.18E-07	3.96E-07	3.75E-07	3.55E-07	3.36E-07	3.18E-07	3.01E-07	2.85E-07
5.50	2.70E-07	2.55E-07	2.42E-07	2.29E-07	2.16E-07	2.05E-07	1.94E-07	1.83E-07	1.73E-07	1.64E-07
5.60	1.55E-07	1.47E-07	1.39E-07	1.31E-07	1.24E-07	1.17E-07	1.11E-07	1.04E-07	9.87E-08	9.32E-08
5.70	8.81E-08	8.32E-08	7.86E-08	7.42E-08	7.01E-08	6.61E-08	6.24E-08	5.89E-08	5.56E-08	5.25E-08
5.80	4.96E-08	4.68E-08	4.41E-08	4.16E-08	3.93E-08	3.70E-08	3.49E-08	3.29E-08	3.11E-08	2.93E-08
5.90	2.76E-08	2.60E-08	2.45E-08	2.31E-08	2.18E-08	2.05E-08	1.93E-08	1.82E-08	1.72E-08	1.62E-08
6.00	1.52E-08	1.43E-08	1.35E-08	1.27E-08	1.20E-08	1.13E-08	1.06E-08	9.98E-09	9.39E-09	8.84E-09
6.10	8.31E-09	7.82E-09	7.36E-09	6.92E-09	6.51E-09	6.12E-09	5.76E-09	5.41E-09	5.09E-09	4.78E-09
6.20	4.50E-09	4.22E-09	3.97E-09	3.73E-09	3.50E-09	3.29E-09	3.09E-09	2.91E-09	2.73E-09	2.56E-09
6.30	2.41E-09	2.26E-09	2.12E-09	1.99E-09	1.87E-09	1.75E-09	1.65E-09	1.54E-09	1.45E-09	1.36E-09
6.40	1.27E-09	1.20E-09	1.12E-09	1.05E-09	9.86E-10	9.25E-10	8.67E-10	8.13E-10	7.62E-10	7.14E-10
6.50	6.69E-10	6.27E-10	5.87E-10	5.50E-10	5.15E-10	4.83E-10	4.52E-10	4.23E-10	3.96E-10	3.71E-10
6.60	3.47E-10	3.25E-10	3.04E-10	2.85E-10	2.67E-10	2.49E-10	2.33E-10	2.18E-10	2.04E-10	1.91E-10
6.70	1.79E-10	1.67E-10	1.56E-10	1.46E-10	1.37E-10	1.28E-10	1.19E-10	1.12E-10	1.04E-10	9.74E-11
6.80	9.10E-11	8.50E-11	7.94E-11	7.42E-11	6.93E-11	6.47E-11	6.04E-11	5.64E-11	5.26E-11	4.91E-11
6.90	4.59E-11	4.28E-11	3.99E-11	3.73E-11	3.48E-11	3.24E-11	3.03E-11	2.82E-11	2.63E-11	2.45E-11
7.00	2.29E-11	2.13E-11	1.99E-11	1.85E-11	1.73E-11	1.61E-11	1.50E-11	1.40E-11	1.30E-11	1.21E-11
7.10	1.13E-11	1.05E-11	9.81E-12	9.13E-12	8.51E-12	7.92E-12	7.37E-12	6.86E-12	6.39E-12	5.94E-12
7.20	5.53E-12	5.15E-12	4.79E-12	4.46E-12	4.14E-12	3.85E-12	3.58E-12	3.33E-12	3.10E-12	2.88E-12
7.30	2.68E-12	2.49E-12	2.31E-12	2.15E-12	2.00E-12	1.86E-12	1.73E-12	1.60E-12	1.49E-12	1.38E-12
7.40	1.28E-12	1.19E-12	1.11E-12	1.03E-12	9.55E-13	8.86E-13	8.23E-13	7.63E-13	7.08E-13	6.57E-13
7.50	6.10E-13	5.66E-13	5.25E-13	4.87E-13	4.51E-13	4.19E-13	3.88E-13	3.60E-13	3.34E-13	3.09E-13

Table 2.1 (Part 4) Solutions to Exponentials: B = exp (-0.5 Asq) where Asq is A squared, or A times A.
Note that 9.67E-01 means 9.67 times 10 to the -1 power.

A	0.00 B	0.01	0.02	0.03	0.04	0.05	0.06	0.07	0.08	0.09
7.60	2.87E-13	2.66E-13	2.46E-13	2.28E-13	2.11E-13	1.96E-13	1.81E-13	1.68E-13	1.56E-13	1.44E-13
7.70	1.33E-13	1.23E-13	1.14E-13	1.06E-13	9.79E-14	9.06E-14	8.39E-14	7.76E-14	7.18E-14	6.64E-14
7.80	6.14E-14	5.68E-14	5.26E-14	4.86E-14	4.49E-14	4.15E-14	3.84E-14	3.55E-14	3.28E-14	3.03E-14
7.90	2.80E-14	2.59E-14	2.39E-14	2.21E-14	2.04E-14	1.89E-14	1.74E-14	1.61E-14	1.48E-14	1.37E-14
8.00	1.27E-14	1.17E-14	1.08E-14	9.95E-15	9.18E-15	8.47E-15	7.82E-15	7.21E-15	6.65E-15	6.13E-15
8.10	5.66E-15	5.22E-15	4.81E-15	4.44E-15	4.09E-15	3.77E-15	3.47E-15	3.20E-15	2.95E-15	2.72E-15
8.20	2.50E-15	2.31E-15	2.13E-15	1.96E-15	1.80E-15	1.66E-15	1.53E-15	1.41E-15	1.30E-15	1.19E-15
8.30	1.10E-15	1.01E-15	9.29E-16	8.55E-16	7.87E-16	7.24E-16	6.66E-16	6.12E-16	5.63E-16	5.18E-16
8.40	4.76E-16	4.38E-16	4.02E-16	3.70E-16	3.40E-16	3.12E-16	2.87E-16	2.64E-16	2.42E-16	2.23E-16
8.50	2.05E-16	1.88E-16	1.73E-16	1.58E-16	1.45E-16	1.34E-16	1.23E-16	1.13E-16	1.03E-16	9.48E-17
8.60	8.70E-17	7.98E-17	7.32E-17	6.72E-17	6.16E-17	5.65E-17	5.18E-17	4.75E-17	4.36E-17	3.99E-17
8.70	3.66E-17	3.36E-17	3.08E-17	2.82E-17	2.58E-17	2.37E-17	2.17E-17	1.99E-17	1.82E-17	1.67E-17
8.80	1.53E-17	1.40E-17	1.28E-17	1.17E-17	1.07E-17	9.82E-18	8.99E-18	8.22E-18	7.53E-18	6.89E-18
8.90	6.30E-18	5.76E-18	5.27E-18	4.82E-18	4.41E-18	4.03E-18	3.69E-18	3.37E-18	3.08E-18	2.82E-18
9.00	2.57E-18	2.35E-18	2.15E-18	1.96E-18	1.79E-18	1.64E-18	1.50E-18	1.37E-18	1.25E-18	1.14E-18
9.10	1.04E-18	9.51E-19	8.68E-19	7.92E-19	7.23E-19	6.60E-19	6.02E-19	5.49E-19	5.01E-19	4.57E-19
9.20	4.17E-19	3.80E-19	3.47E-19	3.16E-19	2.88E-19	2.63E-19	2.40E-19	2.18E-19	1.99E-19	1.81E-19
9.30	1.65E-19	1.51E-19	1.37E-19	1.25E-19	1.14E-19	1.04E-19	9.45E-20	8.60E-20	7.83E-20	7.13E-20
9.40	6.49E-20	5.91E-20	5.38E-20	4.89E-20	4.45E-20	4.05E-20	3.69E-20	3.35E-20	3.05E-20	2.77E-20
9.50	2.52E-20	2.29E-20	2.09E-20	1.90E-20	1.72E-20	1.57E-20	1.42E-20	1.29E-20	1.18E-20	1.07E-20
9.60	9.71E-21	8.82E-21	8.01E-21	7.28E-21	6.61E-21	6.00E-21	5.45E-21	4.95E-21	4.49E-21	4.08E-21
9.70	3.70E-21	3.36E-21	3.05E-21	2.76E-21	2.51E-21	2.27E-21	2.06E-21	1.87E-21	1.70E-21	1.54E-21
9.80	1.40E-21	1.26E-21	1.15E-21	1.04E-21	9.42E-22	8.54E-22	7.74E-22	7.01E-22	6.35E-22	5.75E-22
9.90	5.21E-22	4.72E-22	4.27E-22	3.87E-22	3.50E-22	3.17E-22	2.87E-22	2.60E-22	2.35E-22	2.13E-22
10.00	1.93E-22	1.74E-22	1.58E-22	1.43E-22	1.29E-22	1.17E-22	1.06E-22	9.54E-23	8.63E-23	7.80E-23

Table 2.5 (Part 1) Pasquill-Gifford Dispersion Parameters

x, km	Sigma-y, meters						Sigma-z, meters						Sigma-y times Sigma-z					
	A	B	C	D	E	F	A	B	C	D	E	F	A	B	C	D	E	F
0.01	3.36	2.34	1.47	0.96	0.72	0.48	1.58	1.24	0.91	0.63	0.51	0.36	5.32	2.90	1.33	0.60	0.37	0.17
0.02	6.29	4.42	2.80	1.84	1.37	0.91	3.05	2.37	1.71	1.15	0.92	0.63	19.2	10.5	4.78	2.11	1.26	0.57
0.03	9.08	6.41	4.08	2.68	2.00	1.33	4.47	3.45	2.47	1.63	1.29	0.87	40.6	22.1	10.1	4.37	2.58	1.16
0.04	11.8	8.34	5.33	3.50	2.61	1.74	5.87	4.51	3.22	2.10	1.64	1.10	69.1	37.7	17.2	7.34	4.29	1.91
0.05	14.4	10.2	6.56	4.31	3.22	2.14	7.25	5.56	3.95	2.55	1.98	1.32	104.	56.9	25.9	11.0	6.37	2.82
0.06	17.0	12.1	7.77	5.11	3.81	2.53	8.61	6.59	4.66	2.98	2.31	1.53	146.	79.6	36.2	15.2	8.79	3.88
0.07	19.5	13.9	8.96	5.89	4.40	2.92	9.96	7.61	5.37	3.41	2.62	1.74	194.	106.	48.1	20.1	11.5	5.08
0.08	22.0	15.7	10.1	6.67	4.98	3.31	11.3	8.61	6.07	3.83	2.93	1.94	248.	135.	61.5	25.5	14.6	6.41
0.09	24.4	17.5	11.3	7.44	5.55	3.69	12.6	9.61	6.76	4.24	3.24	2.13	308.	168.	76.4	31.6	18.0	7.88
0.10	26.9	19.3	12.5	8.20	6.12	4.07	13.9	10.6	7.44	4.65	3.53	2.33	375.	204.	92.7	38.1	21.6	9.46
0.11	29.3	21.0	13.6	8.96	6.69	4.45	15.4	11.6	8.12	5.05	3.82	2.51	451.	244.	111.	45.3	25.6	11.2
0.12	31.6	22.7	14.7	9.71	7.25	4.82	16.9	12.6	8.79	5.45	4.10	2.70	535.	286.	130.	52.9	29.8	13.0
0.13	34.0	24.5	15.9	10.5	7.81	5.19	18.4	13.5	9.46	5.84	4.38	2.88	625.	331.	150.	61.1	34.2	14.9
0.14	36.3	26.2	17.0	11.2	8.36	5.56	19.9	14.5	10.1	6.23	4.66	3.06	722.	380.	172.	69.8	38.9	17.0
0.15	38.6	27.9	18.1	11.9	8.91	5.92	21.4	15.5	10.8	6.62	4.93	3.24	826.	431.	195.	79.0	43.9	19.2
0.16	40.9	29.5	19.2	12.7	9.46	6.29	23.0	16.4	11.4	7.00	5.20	3.41	940.	485.	220.	88.7	49.2	21.5
0.17	43.2	31.2	20.3	13.4	10.0	6.65	24.5	17.4	12.1	7.38	5.46	3.58	1.06E+03	543.	246.	98.8	54.6	23.8
0.18	45.5	32.9	21.4	14.1	10.5	7.01	26.1	18.3	12.7	7.76	5.72	3.76	1.19E+03	603.	273.	110.	60.4	26.3
0.19	47.7	34.5	22.5	14.8	11.1	7.37	27.7	19.3	13.4	8.13	5.98	3.93	1.32E+03	666.	302.	121.	66.3	28.9
0.20	50.0	36.2	23.6	15.6	11.6	7.73	29.3	20.2	14.0	8.50	6.24	4.09	1.46E+03	732.	331.	132.	72.5	31.6
0.21	52.2	37.8	24.7	16.3	12.2	8.08	31.0	21.2	14.7	8.87	6.49	4.25	1.62E+03	802.	362.	144.	79.0	34.4
0.22	54.4	39.4	25.8	17.0	12.7	8.44	32.6	22.2	15.3	9.23	6.75	4.41	1.78E+03	876.	395.	157.	85.6	37.2
0.23	56.6	41.0	26.9	17.7	13.2	8.79	34.3	23.2	15.9	9.60	7.00	4.57	1.94E+03	953.	428.	170.	92.5	40.2
0.24	58.8	42.7	27.9	18.4	13.8	9.14	36.0	24.2	16.6	9.96	7.24	4.72	2.12E+03	1.03E+03	463.	183.	99.6	43.2
0.25	61.0	44.3	29.0	19.1	14.3	9.50	37.7	25.2	17.2	10.3	7.49	4.88	2.30E+03	1.12E+03	499.	197.	107.	46.3
0.26	63.2	45.9	30.1	19.8	14.8	9.85	39.6	26.2	17.8	10.7	7.74	5.03	2.50E+03	1.20E+03	536.	212.	115.	49.5
0.27	65.3	47.5	31.1	20.5	15.3	10.2	41.5	27.2	18.5	11.0	7.98	5.18	2.71E+03	1.29E+03	575.	226.	122.	52.8
0.28	67.5	49.0	32.2	21.2	15.9	10.5	43.5	28.2	19.1	11.4	8.22	5.33	2.93E+03	1.38E+03	614.	242.	130.	56.2
0.29	69.6	50.6	33.2	21.9	16.4	10.9	45.5	29.2	19.7	11.7	8.46	5.48	3.16E+03	1.48E+03	655.	257.	139.	59.6
0.30	71.8	52.2	34.3	22.6	16.9	11.2	47.4	30.1	20.3	12.1	8.70	5.62	3.40E+03	1.57E+03	697.	273.	147.	63.2

Table 2.5 (Part 2) Pasquill-Gifford Dispersion Parameters

x, km	Sigma-y, meters						Sigma-z, meters						Sigma-y times Sigma-z					
	A	B	C	D	E	F	A	B	C	D	E	F	A	B	C	D	E	F
0.31	73.9	53.8	35.3	23.3	17.4	11.6	49.7	31.1	20.9	12.4	8.92	5.77	3.67E+03	1.67E+03	740.	289.	155.	66.8
0.32	76.0	55.3	36.4	24.0	17.9	11.9	52.0	32.1	21.6	12.7	9.13	5.92	3.95E+03	1.78E+03	785.	306.	164.	70.5
0.33	78.1	56.9	37.4	24.7	18.4	12.3	54.3	33.1	22.2	13.1	9.35	6.06	4.24E+03	1.88E+03	830.	322.	172.	74.3
0.34	80.2	58.5	38.5	25.4	19.0	12.6	56.6	34.1	22.8	13.4	9.56	6.20	4.54E+03	1.99E+03	877.	340.	181.	78.2
0.35	82.3	60.0	39.5	26.1	19.5	12.9	59.0	35.1	23.4	13.7	9.77	6.35	4.85E+03	2.10E+03	925.	357.	190.	82.2
0.36	84.4	61.5	40.5	26.7	20.0	13.3	61.3	36.1	24.0	14.0	9.98	6.49	5.18E+03	2.22E+03	974.	375.	199.	86.2
0.37	86.5	63.1	41.6	27.4	20.5	13.6	63.8	37.0	24.6	14.3	10.2	6.63	5.52E+03	2.34E+03	1.02E+03	393.	209.	90.3
0.38	88.6	64.6	42.6	28.1	21.0	14.0	66.2	38.0	25.2	14.6	10.4	6.77	5.86E+03	2.46E+03	1.07E+03	412.	218.	94.5
0.39	90.6	66.2	43.6	28.8	21.5	14.3	68.7	39.0	25.8	15.0	10.6	6.91	6.22E+03	2.58E+03	1.13E+03	430.	228.	98.8
0.40	92.7	67.7	44.6	29.5	22.0	14.6	71.2	40.0	26.4	15.3	10.8	7.05	6.60E+03	2.71E+03	1.18E+03	450.	238.	103.
0.41	94.8	69.2	45.5	30.1	22.5	15.0	74.3	41.1	27.0	15.6	11.0	7.19	7.04E+03	2.84E+03	1.24E+03	469.	248.	108.
0.42	96.8	70.7	46.7	30.8	23.0	15.3	77.4	42.2	27.7	15.9	11.2	7.32	7.50E+03	2.98E+03	1.29E+03	489.	258.	112.
0.43	98.9	72.2	47.7	31.5	23.5	15.6	80.6	43.3	28.3	16.2	11.4	7.46	7.97E+03	3.13E+03	1.35E+03	510.	269.	117.
0.44	101.	73.8	48.7	32.1	24.0	16.0	83.9	44.4	28.9	16.5	11.6	7.59	8.47E+03	3.28E+03	1.41E+03	530.	279.	121.
0.45	103.	75.3	49.7	32.8	24.5	16.3	87.2	45.5	29.5	16.8	11.8	7.73	8.98E+03	3.43E+03	1.46E+03	551.	290.	126.
0.46	105.	76.8	50.7	33.5	25.0	16.6	90.6	46.6	30.1	17.1	12.0	7.86	9.51E+03	3.58E+03	1.53E+03	573.	301.	131.
0.47	107.	78.3	51.8	34.2	25.5	17.0	94.0	47.7	30.6	17.4	12.2	8.00	1.01E+04	3.74E+03	1.59E+03	594.	312.	136.
0.48	109.	79.8	52.8	34.8	26.0	17.3	97.5	48.9	31.2	17.7	12.4	8.13	1.06E+04	3.90E+03	1.65E+03	616.	323.	141.
0.49	111.	81.3	53.8	35.5	26.5	17.6	101.	50.0	31.8	18.0	12.6	8.26	1.12E+04	4.06E+03	1.71E+03	639.	334.	146.
0.50	113.	82.8	54.8	36.1	27.0	18.0	105.	51.1	32.4	18.3	12.8	8.40	1.18E+04	4.23E+03	1.78E+03	661.	346.	151.
0.55	123.	90.2	59.8	39.4	29.5	19.6	128.	56.7	35.4	19.8	13.8	9.05	1.58E+04	5.11E+03	2.11E+03	780.	406.	177.
0.60	133.	97.5	64.7	42.7	31.9	21.2	154.	62.4	38.3	21.2	14.7	9.69	2.05E+04	6.08E+03	2.48E+03	906.	469.	206.
0.65	143.	105.	69.6	46.0	34.4	22.9	182.	68.1	41.2	22.6	15.6	10.3	2.60E+04	7.14E+03	2.87E+03	1.04E+03	536.	236.
0.70	152.	112.	74.5	49.2	36.8	24.5	213.	73.9	44.1	24.0	16.5	10.9	3.25E+04	8.28E+03	3.29E+03	1.18E+03	607.	267.
0.75	162.	119.	79.3	52.4	39.2	26.1	247.	79.7	47.0	25.4	17.4	11.5	4.00E+04	9.50E+03	3.73E+03	1.33E+03	681.	299.
0.80	171.	126.	84.1	55.6	41.5	27.6	283.	85.6	49.9	26.8	18.3	12.0	4.85E+04	1.08E+04	4.19E+03	1.49E+03	759.	331.
0.85	181.	133.	88.9	58.7	43.9	29.2	322.	91.5	52.7	28.1	19.1	12.5	5.82E+04	1.22E+04	4.69E+03	1.65E+03	840.	365.
0.90	190.	140.	93.7	61.9	46.3	30.8	363.	97.4	55.5	29.5	20.0	13.0	6.91E+04	1.37E+04	5.20E+03	1.82E+03	924.	400.
0.95	199.	147.	98.4	65.0	48.6	32.3	407.	103.	58.3	30.8	20.8	13.5	8.12E+04	1.52E+04	5.74E+03	2.00E+03	1.01E+03	436.
1.00	209.	154.	103.	68.1	50.9	33.9	454.	109.	61.1	32.1	21.6	14.0	9.47E+04	1.68E+04	6.30E+03	2.19E+03	1.10E+03	473.

Table 2.5 (Part 3) Pasquill-Gifford Dispersion Parameters

x, km	Sigma-y, meters						Sigma-z, meters						Sigma-y times Sigma-z					
	A	B	C	D	E	F	A	B	C	D	E	F	A	B	C	D	E	F
1.05	218.	161.	108.	71.2	53.3	35.4	503.	115.	63.9	33.1	22.3	14.4	1.10E+05	1.86E+04	6.89E+03	2.36E+03	1.19E+03	510.
1.10	227.	168.	112.	74.3	55.6	37.0	555.	121.	66.7	34.1	23.0	14.8	1.26E+05	2.04E+04	7.50E+03	2.54E+03	1.28E+03	548.
1.15	236.	175.	117.	77.4	57.9	38.5	610.	127.	69.5	35.1	23.6	15.2	1.44E+05	2.22E+04	8.14E+03	2.72E+03	1.37E+03	587.
1.20	245.	181.	122.	80.4	60.2	40.0	668.	134.	72.2	36.1	24.3	15.7	1.64E+05	2.42E+04	8.79E+03	2.90E+03	1.46E+03	627.
1.25	254.	188.	126.	83.5	62.4	41.5	728.	140.	75.0	37.1	24.9	16.1	1.85E+05	2.63E+04	9.47E+03	3.09E+03	1.55E+03	667.
1.30	263.	195.	131.	86.5	64.7	43.0	791.	146.	77.7	38.0	25.5	16.5	2.08E+05	2.84E+04	1.02E+04	3.29E+03	1.65E+03	709.
1.35	272.	201.	135.	89.5	67.0	44.5	857.	152.	80.5	38.9	26.1	16.9	2.33E+05	3.06E+04	1.09E+04	3.49E+03	1.75E+03	751.
1.40	281.	208.	140.	92.6	69.2	46.0	925.	158.	83.2	39.9	26.7	17.3	2.60E+05	3.29E+04	1.16E+04	3.69E+03	1.85E+03	795.
1.45	289.	215.	145.	95.6	71.5	47.5	996.	164.	85.9	40.8	27.3	17.6	2.88E+05	3.53E+04	1.24E+04	3.90E+03	1.95E+03	839.
1.50	298.	221.	149.	98.5	73.7	49.0	1071.	171.	88.6	41.7	27.9	18.0	3.19E+05	3.77E+04	1.32E+04	4.11E+03	2.06E+03	884.
1.55	307.	228.	154.	102.	75.9	50.5	1148.	177.	91.3	42.6	28.5	18.4	3.52E+05	4.03E+04	1.40E+04	4.32E+03	2.17E+03	930.
1.60	316.	234.	158.	104.	78.1	52.0	1227.	183.	94.0	43.4	29.1	18.8	3.87E+05	4.29E+04	1.49E+04	4.54E+03	2.27E+03	977.
1.65	324.	241.	163.	107.	80.4	53.5	1310.	189.	96.7	44.3	29.7	19.2	4.25E+05	4.56E+04	1.57E+04	4.76E+03	2.38E+03	1.02E+03
1.70	333.	247.	167.	110.	82.6	54.9	1395.	196.	99.3	45.2	30.2	19.5	4.64E+05	4.84E+04	1.66E+04	4.99E+03	2.50E+03	1.07E+03
1.75	341.	254.	171.	113.	84.8	56.4	1484.	202.	102.	46.0	30.8	19.9	5.06E+05	5.13E+04	1.75E+04	5.22E+03	2.61E+03	1.12E+03
1.80	350.	260.	176.	116.	87.0	57.9	1575.	208.	105.	46.9	31.3	20.2	5.51E+05	5.42E+04	1.84E+04	5.45E+03	2.73E+03	1.17E+03
1.85	358.	267.	180.	119.	89.2	59.3	1669.	215.	107.	47.7	31.9	20.6	5.98E+05	5.72E+04	1.93E+04	5.69E+03	2.84E+03	1.22E+03
1.90	367.	273.	185.	122.	91.3	60.8	1766.	221.	110.	48.5	32.4	20.9	6.48E+05	6.04E+04	2.03E+04	5.93E+03	2.96E+03	1.27E+03
1.95	375.	279.	189.	125.	93.5	62.2	1866.	227.	113.	49.3	33.0	21.3	7.00E+05	6.36E+04	2.13E+04	6.17E+03	3.08E+03	1.32E+03
2.00	384.	286.	193.	128.	95.7	63.7	1968.	234.	115.	50.2	33.5	21.6	7.55E+05	6.68E+04	2.23E+04	6.42E+03	3.20E+03	1.38E+03
2.10	400.	298.	202.	134.	100.	66.6	2182.	247.	121.	51.8	34.4	22.2	8.74E+05	7.36E+04	2.44E+04	6.92E+03	3.44E+03	1.48E+03
2.20	417.	311.	211.	139.	104.	69.4	2408.	260.	126.	53.3	35.4	22.8	1.00E+06	8.07E+04	2.65E+04	7.44E+03	3.69E+03	1.58E+03
2.30	433.	324.	220.	145.	109.	72.3	2646.	273.	131.	54.9	36.3	23.3	1.15E+06	8.82E+04	2.87E+04	7.97E+03	3.94E+03	1.69E+03
2.40	450.	336.	228.	151.	113.	75.1	2895.	286.	136.	56.4	37.2	23.9	1.30E+06	9.59E+04	3.11E+04	8.51E+03	4.20E+03	1.79E+03
2.50	466.	348.	237.	157.	117.	77.9	3156.	299.	141.	57.9	38.0	24.4	1.47E+06	1.04E+05	3.35E+04	9.07E+03	4.46E+03	1.90E+03
2.60	482.	361.	245.	162.	121.	80.8	3430.	312.	147.	59.4	38.9	25.0	1.65E+06	1.12E+05	3.59E+04	9.64E+03	4.72E+03	2.02E+03
2.70	498.	373.	254.	168.	126.	83.6	3715.	325.	152.	60.8	39.8	25.5	1.85E+06	1.21E+05	3.85E+04	1.02E+04	4.99E+03	2.13E+03
2.80	515.	385.	262.	173.	130.	86.4	4012.	338.	157.	62.3	40.6	26.0	2.06E+06	1.30E+05	4.11E+04	1.08E+04	5.27E+03	2.24E+03
2.90	530.	397.	271.	179.	134.	89.1	4321.	351.	162.	63.7	41.4	26.5	2.29E+06	1.40E+05	4.38E+04	1.14E+04	5.55E+03	2.36E+03
3.00	546.	409.	279.	185.	138.	91.9	4643.	365.	167.	65.1	42.2	27.0	2.54E+06	1.49E+05	4.66E+04	1.20E+04	5.83E+03	2.48E+03

Table 2.5 (Part 4) Pasquill-Gifford Dispersion Parameters

x, km	Sigma-y, meters						Sigma-z, meters						Sigma-y times Sigma-z					
	A	B	C	D	E	F	A	B	C	D	E	F	A	B	C	D	E	F
3.10	562.	421.	287.	190.	142.	94.7	4977.	378.	172.	66.4	43.0	27.4	2.80E+06	1.59E+05	4.95E+04	1.26E+04	6.12E+03	2.59E+03
3.20	578.	433.	296.	196.	146.	97.4	5000.	392.	177.	67.7	43.8	27.8	2.89E+06	1.70E+05	5.24E+04	1.33E+04	6.41E+03	2.71E+03
3.30	594.	445.	304.	201.	151.	100.	5000.	405.	182.	69.0	44.6	28.2	2.97E+06	1.80E+05	5.54E+04	1.39E+04	6.71E+03	2.83E+03
3.40	609.	457.	312.	207.	155.	103.	5000.	419.	187.	70.2	45.4	28.6	3.05E+06	1.91E+05	5.85E+04	1.45E+04	7.01E+03	2.94E+03
3.50	625.	469.	321.	212.	159.	106.	5000.	432.	192.	71.5	46.1	29.0	3.12E+06	2.03E+05	6.16E+04	1.52E+04	7.32E+03	3.06E+03
3.60	640.	481.	329.	218.	163.	108.	5000.	446.	197.	72.7	46.9	29.4	3.20E+06	2.14E+05	6.49E+04	1.58E+04	7.63E+03	3.18E+03
3.70	656.	492.	337.	223.	167.	111.	5000.	459.	202.	73.9	47.6	29.7	3.28E+06	2.26E+05	6.82E+04	1.65E+04	7.94E+03	3.30E+03
3.80	671.	504.	345.	229.	171.	114.	5000.	473.	207.	75.1	48.3	30.1	3.35E+06	2.38E+05	7.16E+04	1.72E+04	8.26E+03	3.42E+03
3.90	686.	516.	353.	234.	175.	116.	5000.	486.	212.	76.3	49.1	30.5	3.43E+06	2.51E+05	7.50E+04	1.79E+04	8.59E+03	3.55E+03
4.00	701.	527.	361.	239.	179.	119.	5000.	500.	217.	77.5	49.8	30.8	3.51E+06	2.64E+05	7.85E+04	1.85E+04	8.91E+03	3.67E+03
4.10	716.	539.	370.	245.	183.	122.	5000.	514.	222.	78.7	50.4	31.2	3.58E+06	2.77E+05	8.21E+04	1.92E+04	9.23E+03	3.80E+03
4.20	732.	550.	378.	250.	187.	125.	5000.	528.	227.	79.8	51.0	31.5	3.66E+06	2.90E+05	8.58E+04	2.00E+04	9.54E+03	3.93E+03
4.30	747.	562.	386.	255.	191.	127.	5000.	541.	232.	81.0	51.6	31.9	3.73E+06	3.04E+05	8.95E+04	2.07E+04	9.86E+03	4.06E+03
4.40	762.	573.	394.	261.	195.	130.	5000.	555.	237.	82.1	52.2	32.2	3.81E+06	3.18E+05	9.34E+04	2.14E+04	1.02E+04	4.19E+03
4.50	777.	585.	402.	266.	199.	133.	5000.	569.	242.	83.2	52.8	32.6	3.88E+06	3.33E+05	9.72E+04	2.21E+04	1.05E+04	4.32E+03
4.60	791.	596.	410.	271.	203.	135.	5000.	583.	247.	84.3	53.4	32.9	3.96E+06	3.48E+05	1.01E+05	2.29E+04	1.08E+04	4.45E+03
4.70	806.	608.	418.	277.	207.	138.	5000.	597.	252.	85.4	54.0	33.2	4.03E+06	3.63E+05	1.05E+05	2.36E+04	1.12E+04	4.58E+03
4.80	821.	619.	426.	282.	211.	140.	5000.	611.	257.	86.5	54.6	33.6	4.11E+06	3.78E+05	1.09E+05	2.44E+04	1.15E+04	4.71E+03
4.90	836.	630.	434.	287.	215.	143.	5000.	625.	262.	87.6	55.1	33.9	4.18E+06	3.94E+05	1.13E+05	2.52E+04	1.19E+04	4.85E+03
5.00	851.	641.	442.	292.	219.	146.	5000.	639.	266.	88.7	55.7	34.2	4.25E+06	4.10E+05	1.18E+05	2.59E+04	1.22E+04	4.98E+03
5.50	923.	697.	481.	319.	238.	159.	5000.	709.	291.	94.0	58.5	35.8	4.62E+06	4.95E+05	1.40E+05	2.99E+04	1.39E+04	5.67E+03
6.00	995.	752.	520.	344.	258.	172.	5000.	780.	315.	99.0	61.1	37.2	4.98E+06	5.87E+05	1.64E+05	3.41E+04	1.57E+04	6.39E+03
6.50	1066.	807.	559.	370.	277.	184.	5000.	852.	339.	104.	63.6	38.6	5.33E+06	6.88E+05	1.89E+05	3.85E+04	1.76E+04	7.12E+03
7.00	1136.	861.	597.	395.	296.	197.	5000.	924.	362.	109.	66.0	40.0	5.68E+06	7.96E+05	2.16E+05	4.30E+04	1.95E+04	7.88E+03
7.50	1205.	914.	635.	421.	315.	210.	5000.	997.	386.	113.	68.4	41.2	6.03E+06	9.11E+05	2.45E+05	4.77E+04	2.15E+04	8.62E+03
8.00	1274.	967.	672.	446.	333.	222.	5000.	1070.	410.	118.	70.6	42.3	6.37E+06	1.03E+06	2.75E+05	5.25E+04	2.36E+04	9.38E+03
8.50	1342.	1020.	710.	470.	352.	234.	5000.	1144.	433.	122.	72.8	43.4	6.71E+06	1.17E+06	3.07E+05	5.75E+04	2.56E+04	1.02E+04
9.00	1409.	1071.	747.	495.	370.	247.	5000.	1218.	456.	127.	75.0	44.4	7.04E+06	1.30E+06	3.41E+05	6.26E+04	2.78E+04	1.09E+04
9.50	1475.	1123.	784.	519.	389.	259.	5000.	1292.	479.	131.	77.0	45.4	7.38E+06	1.45E+06	3.76E+05	6.79E+04	3.00E+04	1.17E+04
10.00	1541.	1174.	820.	544.	407.	271.	5000.	1367.	502.	135.	79.1	46.4	7.71E+06	1.60E+06	4.12E+05	7.33E+04	3.22E+04	1.26E+04

Table 2.5 (Part 5) Pasquill-Gifford Dispersion Parameters

x, km	Sigma-y, meters						Sigma-z, meters						Sigma-y times Sigma-z					
	A	B	C	D	E	F	A	B	C	D	E	F	A	B	C	D	E	F
10.50	1607.	1225.	856.	568.	425.	283.	5000.	1442.	525.	139.	80.9	47.3	8.03E+06	1.77E+06	4.50E+05	7.87E+04	3.44E+04	1.34E+04
11.00	1671.	1275.	893.	592.	443.	295.	5000.	1518.	548.	142.	82.7	48.2	8.36E+06	1.93E+06	4.89E+05	8.42E+04	3.66E+04	1.42E+04
11.50	1736.	1325.	929.	616.	461.	307.	5000.	1593.	571.	146.	84.4	49.1	8.68E+06	2.11E+06	5.30E+05	8.99E+04	3.89E+04	1.51E+04
12.00	1800.	1375.	964.	639.	479.	319.	5000.	1670.	593.	150.	86.1	50.0	9.00E+06	2.30E+06	5.72E+05	9.56E+04	4.12E+04	1.59E+04
12.50	1863.	1424.	1000.	663.	496.	330.	5000.	1746.	616.	153.	87.8	50.9	9.32E+06	2.49E+06	6.16E+05	1.01E+05	4.36E+04	1.68E+04
13.00	1926.	1473.	1035.	686.	514.	342.	5000.	1823.	639.	156.	89.4	51.7	9.63E+06	2.68E+06	6.61E+05	1.07E+05	4.59E+04	1.77E+04
13.50	1989.	1522.	1070.	710.	531.	354.	5000.	1900.	661.	160.	91.0	52.5	9.94E+06	2.89E+06	7.08E+05	1.13E+05	4.83E+04	1.86E+04
14.00	2051.	1570.	1105.	733.	549.	365.	5000.	1977.	683.	163.	92.5	53.3	1.03E+07	3.10E+06	7.55E+05	1.20E+05	5.08E+04	1.95E+04
14.50	2113.	1618.	1140.	756.	566.	377.	5000.	2055.	706.	166.	94.1	54.1	1.06E+07	3.33E+06	8.05E+05	1.26E+05	5.32E+04	2.04E+04
15.00	2174.	1666.	1175.	779.	583.	388.	5000.	2133.	728.	170.	95.6	54.9	1.09E+07	3.55E+06	8.55E+05	1.32E+05	5.57E+04	2.13E+04
15.50	2235.	1714.	1210.	802.	601.	400.	5000.	2211.	750.	173.	97.0	55.5	1.12E+07	3.79E+06	9.07E+05	1.39E+05	5.83E+04	2.22E+04
16.00	2296.	1761.	1244.	825.	618.	411.	5000.	2289.	772.	176.	98.5	56.1	1.15E+07	4.03E+06	9.60E+05	1.45E+05	6.08E+04	2.31E+04
16.50	2356.	1809.	1278.	848.	635.	423.	5000.	2368.	794.	179.	99.9	56.6	1.18E+07	4.28E+06	1.02E+06	1.52E+05	6.34E+04	2.39E+04
17.00	2416.	1855.	1312.	871.	652.	434.	5000.	2447.	816.	182.	101.	57.2	1.21E+07	4.54E+06	1.07E+06	1.59E+05	6.60E+04	2.48E+04
17.50	2475.	1902.	1346.	893.	669.	445.	5000.	2526.	838.	185.	103.	57.7	1.24E+07	4.80E+06	1.13E+06	1.65E+05	6.87E+04	2.57E+04
18.00	2535.	1949.	1380.	916.	686.	456.	5000.	2605.	860.	188.	104.	58.3	1.27E+07	5.08E+06	1.19E+06	1.72E+05	7.13E+04	2.66E+04
18.50	2594.	1995.	1414.	938.	702.	468.	5000.	2684.	882.	191.	105.	58.8	1.30E+07	5.36E+06	1.25E+06	1.79E+05	7.40E+04	2.75E+04
19.00	2653.	2041.	1448.	960.	719.	479.	5000.	2764.	904.	194.	107.	59.3	1.33E+07	5.64E+06	1.31E+06	1.86E+05	7.67E+04	2.84E+04
19.50	2711.	2087.	1481.	983.	736.	490.	5000.	2844.	925.	197.	108.	59.8	1.36E+07	5.93E+06	1.37E+06	1.93E+05	7.95E+04	2.93E+04
20.00	2769.	2133.	1515.	1005.	752.	501.	5000.	2924.	947.	200.	109.	60.3	1.38E+07	6.24E+06	1.43E+06	2.01E+05	8.22E+04	3.02E+04
21.00	2885.	2223.	1581.	1049.	785.	523.	5000.	3085.	990.	205.	111.	61.3	1.44E+07	6.86E+06	1.57E+06	2.15E+05	8.74E+04	3.20E+04
22.00	2999.	2313.	1647.	1093.	818.	545.	5000.	3246.	1033.	211.	113.	62.2	1.50E+07	7.51E+06	1.70E+06	2.30E+05	9.27E+04	3.39E+04
23.00	3113.	2403.	1713.	1136.	851.	567.	5000.	3409.	1076.	216.	115.	63.1	1.56E+07	8.19E+06	1.84E+06	2.46E+05	9.80E+04	3.58E+04
24.00	3225.	2491.	1778.	1180.	883.	588.	5000.	3571.	1119.	221.	117.	64.0	1.61E+07	8.90E+06	1.99E+06	2.61E+05	1.03E+05	3.76E+04
25.00	3337.	2580.	1843.	1223.	916.	610.	5000.	3735.	1161.	227.	119.	64.9	1.67E+07	9.63E+06	2.14E+06	2.77E+05	1.09E+05	3.95E+04
26.00	3448.	2667.	1907.	1266.	948.	631.	5000.	3899.	1204.	232.	121.	65.7	1.72E+07	1.04E+07	2.30E+06	2.93E+05	1.14E+05	4.15E+04
27.00	3558.	2754.	1971.	1308.	980.	652.	5000.	4064.	1246.	237.	122.	66.5	1.78E+07	1.12E+07	2.46E+06	3.10E+05	1.20E+05	4.34E+04
28.00	3667.	2840.	2035.	1351.	1011.	674.	5000.	4230.	1288.	242.	124.	67.3	1.83E+07	1.20E+07	2.62E+06	3.26E+05	1.25E+05	4.53E+04
29.00	3775.	2926.	2099.	1393.	1043.	695.	5000.	4396.	1330.	246.	126.	68.1	1.89E+07	1.29E+07	2.79E+06	3.43E+05	1.31E+05	4.73E+04
30.00	3883.	3011.	2162.	1435.	1075.	716.	5000.	4562.	1372.	251.	127.	68.8	1.94E+07	1.37E+07	2.97E+06	3.60E+05	1.37E+05	4.93E+04

Table 2.5 (Part 6) Pasquill-Gifford Dispersion Parameters

x, km	Sigma-y, meters						Sigma-z, meters						Sigma-y times Sigma-z					
	A	B	C	D	E	F	A	B	C	D	E	F	A	B	C	D	E	F
31.00	3990.	3096.	2225.	1477.	1106.	736.	5000.	4729.	1414.	255.	129.	69.5	1.99E+07	1.46E+07	3.15E+06	3.77E+05	1.43E+05	5.12E+04
32.00	4096.	3180.	2287.	1518.	1137.	757.	5000.	4897.	1456.	260.	130.	70.1	2.05E+07	1.56E+07	3.33E+06	3.94E+05	1.48E+05	5.31E+04
33.00	4202.	3264.	2350.	1560.	1168.	778.	5000.	5000.	1497.	264.	132.	70.7	2.10E+07	1.63E+07	3.52E+06	4.11E+05	1.54E+05	5.50E+04
34.00	4306.	3347.	2412.	1601.	1199.	798.	5000.	5000.	1539.	268.	133.	71.2	2.15E+07	1.67E+07	3.71E+06	4.29E+05	1.60E+05	5.69E+04
35.00	4411.	3430.	2474.	1642.	1230.	819.	5000.	5000.	1580.	272.	135.	71.8	2.21E+07	1.72E+07	3.91E+06	4.46E+05	1.66E+05	5.88E+04
36.00	4514.	3513.	2535.	1683.	1260.	839.	5000.	5000.	1621.	276.	136.	72.4	2.26E+07	1.76E+07	4.11E+06	4.64E+05	1.72E+05	6.07E+04
37.00	4617.	3595.	2596.	1724.	1291.	860.	5000.	5000.	1662.	280.	138.	72.9	2.31E+07	1.80E+07	4.32E+06	4.82E+05	1.78E+05	6.27E+04
38.00	4720.	3676.	2657.	1764.	1321.	880.	5000.	5000.	1703.	283.	139.	73.4	2.36E+07	1.84E+07	4.53E+06	5.00E+05	1.84E+05	6.46E+04
39.00	4822.	3758.	2718.	1805.	1352.	900.	5000.	5000.	1744.	287.	141.	74.0	2.41E+07	1.88E+07	4.74E+06	5.18E+05	1.90E+05	6.66E+04
40.00	4923.	3838.	2779.	1845.	1382.	920.	5000.	5000.	1785.	291.	142.	74.5	2.46E+07	1.92E+07	4.96E+06	5.37E+05	1.96E+05	6.85E+04
41.00	5024.	3919.	2839.	1885.	1412.	940.	5000.	5000.	1826.	295.	143.	75.0	2.51E+07	1.96E+07	5.18E+06	5.56E+05	2.02E+05	7.05E+04
42.00	5124.	3999.	2899.	1925.	1442.	960.	5000.	5000.	1867.	298.	144.	75.5	2.56E+07	2.00E+07	5.41E+06	5.74E+05	2.07E+05	7.25E+04
43.00	5224.	4079.	2959.	1965.	1472.	980.	5000.	5000.	1907.	302.	145.	76.0	2.61E+07	2.04E+07	5.64E+06	5.93E+05	2.13E+05	7.45E+04
44.00	5323.	4158.	3019.	2005.	1501.	1000.	5000.	5000.	1948.	306.	146.	76.5	2.66E+07	2.08E+07	5.88E+06	6.12E+05	2.19E+05	7.65E+04
45.00	5422.	4237.	3078.	2044.	1531.	1020.	5000.	5000.	1988.	309.	147.	76.9	2.71E+07	2.12E+07	6.12E+06	6.32E+05	2.25E+05	7.84E+04
46.00	5520.	4316.	3138.	2084.	1560.	1039.	5000.	5000.	2028.	313.	148.	77.4	2.76E+07	2.16E+07	6.36E+06	6.51E+05	2.31E+05	8.04E+04
47.00	5618.	4394.	3197.	2123.	1590.	1059.	5000.	5000.	2069.	316.	149.	77.9	2.81E+07	2.20E+07	6.61E+06	6.71E+05	2.37E+05	8.24E+04
48.00	5715.	4472.	3256.	2162.	1619.	1078.	5000.	5000.	2109.	319.	150.	78.3	2.86E+07	2.24E+07	6.87E+06	6.91E+05	2.42E+05	8.45E+04
49.00	5812.	4550.	3315.	2201.	1649.	1098.	5000.	5000.	2149.	323.	151.	78.8	2.91E+07	2.28E+07	7.12E+06	7.11E+05	2.48E+05	8.65E+04
50.00	5909.	4627.	3373.	2240.	1678.	1117.	5000.	5000.	2189.	326.	152.	79.2	2.95E+07	2.31E+07	7.38E+06	7.31E+05	2.54E+05	8.85E+04
51.00	6005.	4705.	3431.	2279.	1707.	1137.	5000.	5000.	2229.	330.	152.	79.6	3.00E+07	2.35E+07	7.65E+06	7.51E+05	2.60E+05	9.05E+04
52.00	6101.	4781.	3490.	2317.	1736.	1156.	5000.	5000.	2269.	333.	153.	80.0	3.05E+07	2.39E+07	7.92E+06	7.71E+05	2.66E+05	9.25E+04
53.00	6196.	4858.	3548.	2356.	1765.	1175.	5000.	5000.	2309.	336.	154.	80.5	3.10E+07	2.43E+07	8.19E+06	7.92E+05	2.72E+05	9.46E+04
54.00	6291.	4934.	3606.	2395.	1794.	1195.	5000.	5000.	2349.	339.	155.	80.9	3.15E+07	2.47E+07	8.47E+06	8.12E+05	2.78E+05	9.66E+04
55.00	6385.	5010.	3663.	2433.	1822.	1214.	5000.	5000.	2389.	343.	156.	81.3	3.19E+07	2.51E+07	8.75E+06	8.33E+05	2.84E+05	9.87E+04
56.00	6479.	5086.	3721.	2471.	1851.	1233.	5000.	5000.	2428.	346.	157.	81.7	3.24E+07	2.54E+07	9.04E+06	8.54E+05	2.90E+05	1.01E+05
57.00	6573.	5161.	3778.	2509.	1880.	1252.	5000.	5000.	2468.	349.	158.	82.1	3.29E+07	2.58E+07	9.32E+06	8.75E+05	2.96E+05	1.03E+05
58.00	6666.	5237.	3835.	2547.	1908.	1271.	5000.	5000.	2508.	352.	158.	82.5	3.33E+07	2.62E+07	9.62E+06	8.97E+05	3.02E+05	1.05E+05
59.00	6759.	5311.	3892.	2585.	1936.	1290.	5000.	5000.	2547.	355.	159.	82.9	3.38E+07	2.66E+07	9.91E+06	9.18E+05	3.08E+05	1.07E+05
60.00	6852.	5386.	3949.	2623.	1965.	1309.	5000.	5000.	2587.	358.	160.	83.3	3.43E+07	2.69E+07	1.02E+07	9.39E+05	3.14E+05	1.09E+05

Table 2.5   (Part 7)   Pasquill-Gifford Dispersion Parameters

x, km	Sigma-y, meters						Sigma-z, meters						Sigma-y times Sigma-z					
	A	B	C	D	E	F	A	B	C	D	E	F	A	B	C	D	E	F
61.00	6944.	5461.	4006.	2661.	1993.	1328.	5000.	5000.	2626.	361.	161.	83.6	3.47E+07	2.73E+07	1.05E+07	9.61E+05	3.20E+05	1.11E+05
62.00	7036.	5535.	4063.	2698.	2021.	1346.	5000.	5000.	2665.	364.	162.	83.9	3.52E+07	2.77E+07	1.08E+07	9.83E+05	3.26E+05	1.13E+05
63.00	7127.	5609.	4119.	2736.	2049.	1365.	5000.	5000.	2705.	367.	162.	84.1	3.56E+07	2.80E+07	1.11E+07	1.00E+06	3.33E+05	1.15E+05
64.00	7218.	5682.	4175.	2774.	2078.	1384.	5000.	5000.	2744.	370.	163.	84.4	3.61E+07	2.84E+07	1.15E+07	1.03E+06	3.39E+05	1.17E+05
65.00	7309.	5756.	4232.	2811.	2106.	1402.	5000.	5000.	2783.	373.	164.	84.7	3.65E+07	2.88E+07	1.18E+07	1.05E+06	3.45E+05	1.19E+05
66.00	7400.	5829.	4288.	2848.	2133.	1421.	5000.	5000.	2822.	376.	165.	85.0	3.70E+07	2.91E+07	1.21E+07	1.07E+06	3.51E+05	1.21E+05
67.00	7490.	5902.	4344.	2885.	2161.	1440.	5000.	5000.	2861.	379.	165.	85.3	3.75E+07	2.95E+07	1.24E+07	1.09E+06	3.57E+05	1.23E+05
68.00	7580.	5975.	4399.	2922.	2189.	1458.	5000.	5000.	2900.	382.	166.	85.5	3.79E+07	2.99E+07	1.28E+07	1.12E+06	3.63E+05	1.25E+05
69.00	7670.	6047.	4455.	2959.	2217.	1477.	5000.	5000.	2939.	385.	167.	85.8	3.83E+07	3.02E+07	1.31E+07	1.14E+06	3.70E+05	1.27E+05
70.00	7759.	6120.	4511.	2996.	2244.	1495.	5000.	5000.	2978.	388.	167.	86.1	3.88E+07	3.06E+07	1.34E+07	1.16E+06	3.76E+05	1.29E+05
71.00	7848.	6192.	4566.	3033.	2272.	1513.	5000.	5000.	3017.	390.	168.	86.4	3.92E+07	3.10E+07	1.38E+07	1.18E+06	3.82E+05	1.31E+05
72.00	7936.	6264.	4621.	3070.	2300.	1532.	5000.	5000.	3056.	393.	169.	86.6	3.97E+07	3.13E+07	1.41E+07	1.21E+06	3.88E+05	1.33E+05
73.00	8025.	6335.	4676.	3107.	2327.	1550.	5000.	5000.	3095.	396.	169.	86.9	4.01E+07	3.17E+07	1.45E+07	1.23E+06	3.94E+05	1.35E+05
74.00	8113.	6407.	4731.	3143.	2355.	1568.	5000.	5000.	3133.	399.	170.	87.1	4.06E+07	3.20E+07	1.48E+07	1.25E+06	4.01E+05	1.37E+05
75.00	8201.	6478.	4786.	3180.	2382.	1587.	5000.	5000.	3172.	401.	171.	87.4	4.10E+07	3.24E+07	1.52E+07	1.28E+06	4.07E+05	1.39E+05
76.00	8288.	6549.	4841.	3216.	2409.	1605.	5000.	5000.	3211.	404.	172.	87.6	4.14E+07	3.27E+07	1.55E+07	1.30E+06	4.13E+05	1.41E+05
77.00	8375.	6620.	4896.	3252.	2436.	1623.	5000.	5000.	3249.	407.	172.	87.9	4.19E+07	3.31E+07	1.59E+07	1.32E+06	4.20E+05	1.43E+05
78.00	8462.	6691.	4950.	3289.	2464.	1641.	5000.	5000.	3288.	410.	173.	88.1	4.23E+07	3.35E+07	1.63E+07	1.35E+06	4.26E+05	1.45E+05
79.00	8549.	6761.	5005.	3325.	2491.	1659.	5000.	5000.	3327.	412.	174.	88.4	4.27E+07	3.38E+07	1.66E+07	1.37E+06	4.32E+05	1.47E+05
80.00	8635.	6831.	5059.	3361.	2518.	1677.	5000.	5000.	3365.	415.	174.	88.6	4.32E+07	3.42E+07	1.70E+07	1.39E+06	4.38E+05	1.49E+05
81.00	8721.	6902.	5113.	3397.	2545.	1695.	5000.	5000.	3404.	418.	175.	88.9	4.36E+07	3.45E+07	1.74E+07	1.42E+06	4.45E+05	1.51E+05
82.00	8807.	6971.	5167.	3433.	2572.	1713.	5000.	5000.	3442.	420.	175.	89.1	4.40E+07	3.49E+07	1.78E+07	1.44E+06	4.51E+05	1.53E+05
83.00	8893.	7041.	5221.	3469.	2599.	1731.	5000.	5000.	3480.	423.	176.	89.3	4.45E+07	3.52E+07	1.82E+07	1.47E+06	4.58E+05	1.55E+05
84.00	8978.	7111.	5275.	3505.	2626.	1749.	5000.	5000.	3519.	425.	177.	89.6	4.49E+07	3.56E+07	1.86E+07	1.49E+06	4.64E+05	1.57E+05
85.00	9063.	7180.	5329.	3541.	2652.	1767.	5000.	5000.	3557.	428.	177.	89.8	4.53E+07	3.59E+07	1.90E+07	1.52E+06	4.70E+05	1.59E+05
86.00	9148.	7249.	5383.	3576.	2679.	1785.	5000.	5000.	3595.	431.	178.	90.0	4.57E+07	3.62E+07	1.94E+07	1.54E+06	4.77E+05	1.61E+05
87.00	9233.	7318.	5436.	3612.	2706.	1802.	5000.	5000.	3633.	433.	179.	90.3	4.62E+07	3.66E+07	1.98E+07	1.56E+06	4.83E+05	1.63E+05
88.00	9317.	7387.	5490.	3648.	2732.	1820.	5000.	5000.	3672.	436.	179.	90.5	4.66E+07	3.69E+07	2.02E+07	1.59E+06	4.89E+05	1.65E+05
89.00	9401.	7456.	5543.	3683.	2759.	1838.	5000.	5000.	3710.	438.	180.	90.7	4.70E+07	3.73E+07	2.06E+07	1.61E+06	4.96E+05	1.67E+05
90.00	9485.	7524.	5596.	3718.	2786.	1856.	5000.	5000.	3748.	441.	180.	90.9	4.74E+07	3.76E+07	2.10E+07	1.64E+06	5.02E+05	1.69E+05

Table 2.5 (Part 8) Pasquill-Gifford Dispersion Parameters

x, km	Sigma-y, meters						Sigma-z, meters						Sigma-y times Sigma-z					
	A	B	C	D	E	F	A	B	C	D	E	F	A	B	C	D	E	F
91.00	9569.	7593.	5649.	3754.	2812.	1873.	5000.	5000.	3786.	443.	181.	91.1	4.78E+07	3.80E+07	2.14E+07	1.66E+06	5.09E+05	1.71E+05
92.00	9652.	7661.	5702.	3789.	2839.	1891.	5000.	5000.	3824.	446.	182.	91.4	4.83E+07	3.83E+07	2.18E+07	1.69E+06	5.15E+05	1.73E+05
93.00	9736.	7729.	5755.	3824.	2865.	1908.	5000.	5000.	3862.	448.	182.	91.6	4.87E+07	3.86E+07	2.22E+07	1.71E+06	5.22E+05	1.75E+05
94.00	9818.	7797.	5808.	3860.	2891.	1926.	5000.	5000.	3900.	451.	183.	91.8	4.91E+07	3.90E+07	2.27E+07	1.74E+06	5.28E+05	1.77E+05
95.00	9901.	7865.	5861.	3895.	2918.	1944.	5000.	5000.	3938.	453.	183.	92.0	4.95E+07	3.93E+07	2.31E+07	1.76E+06	5.35E+05	1.79E+05
96.00	9984.	7932.	5914.	3930.	2944.	1961.	5000.	5000.	3976.	455.	184.	92.2	4.99E+07	3.97E+07	2.35E+07	1.79E+06	5.41E+05	1.81E+05
97.00	10066.	8000.	5966.	3965.	2970.	1979.	5000.	5000.	4014.	458.	184.	92.4	5.03E+07	4.00E+07	2.39E+07	1.82E+06	5.48E+05	1.83E+05
98.00	10148.	8067.	6019.	4000.	2996.	1996.	5000.	5000.	4051.	460.	185.	92.6	5.07E+07	4.03E+07	2.44E+07	1.84E+06	5.54E+05	1.85E+05
99.00	10230.	8134.	6071.	4034.	3022.	2013.	5000.	5000.	4089.	463.	185.	92.8	5.11E+07	4.07E+07	2.48E+07	1.87E+06	5.61E+05	1.87E+05
100.00	10312.	8201.	6124.	4069.	3049.	2031.	5000.	5000.	4127.	465.	186.	93.0	5.16E+07	4.10E+07	2.53E+07	1.89E+06	5.67E+05	1.89E+05

CHAPTER 3.

EFFECTIVE HEIGHT OF EMISSION

An important consideration in making dispersion estimates is to use the proper effective height of emission above the ground. Although this height may be the physical height of the stack, because of density differences between stack gases and the ambient air, there is likely to be some plume rise following release.

This plume rise may be due to the momentum of the release, the mass of gaseous release times the exit velocity. However, the effects of buoyant release, that due to lower density of plume effluents due to higher temperatures, is more likely to produce a greater effect. A rough rule-of-thumb is that if the exit temperature is about 10 to 15 degrees (K or C) higher than the air temperature, buoyant rise will be greater than that due to momentum. Also, the effects of momentum will be dissipated within 30 to 40 seconds after release. The effects of buoyancy will persist until sufficient ambient air is entrained into the plume to lower its temperature to that of the ambient air. Depending upon the degree of turbulence, the effect of buoyancy is likely to take place over three or four minutes.

3.1 Stack-Tip Downwash

Another effect, that of stack-tip downwash, is due to the regions of low pressure that form in the lee of a stack. The effect is that a cylindrical eddy with vertical axis forms in the low pressure region downwind of the stack and is shed from the stack and travels downwind in the prevailing flow. Immediately the low pressure region begins to be reestablished and another eddy rotating in the opposite direction forms and is shed off into the flow. These are called Karmen vortices after the fluid dynamacist Theodore von Karmen. If the effluent leaving the stack is not clearing the stack well, a part of the bottom of the plume may be caught in the low pressure in the top of these eddies and will be lowered somewhat. A mean correction for this effect was suggested by Briggs (1973, p 3). The correction is to subtract an increment from the physical height of the stack if the ratio of exit velocity to wind speed at stack top, v/u, is less than 1.5. The corrected height, h', is:

$$h' \ = \ h + 2 \, [(v/u_h) - 1.5] \tag{3.1}$$

If v/u is greater than 1.5, it is assumed that there is no stack-tip downwash and the corrected height is the same as the physical height.

$$h' \ = \ h \tag{3.2}$$

Note that the maximum correction is 3 d.

3.2 Buoyancy Flux

Most dispersion models calculate both buoyant and momentum rise and then assume that the dominant mechanism is the one producing the higher rise. To calculate buoyant rise, the buoyancy flux must first be calculated using eq. 12, p 63 of Briggs (1975):

$$F \quad = \quad g \, v \, d^2 \, \Delta T \,/(4 \, T_s) \qquad (3.3)$$

where

	F	buoyancy flux, $m^4 \, s^{-3}$
	g	acceleration of gravity, $9.8 \, m \, s^{-2}$
	v	stack gas exit velocity, m/s
	d	top inside stack diameter, m
	ΔT	stack gas temperature minus ambient air temperature, K
	T_s	stack gas temperature, K

3.3 Final Rise for Unstable-Neutral Conditions

3.3.1 Unstable-Neutral Buoyant Rise

We will first consider buoyant rise for unstable and neutral conditions. The plume rise equations that are generally used are largely empirical, having been derived by fitting lines to graphs of plotted data. Therefore there is a different equation used for buoyancy flux less than 55 and another equation for buoyancy flux more than 55. This author speculates that this may be due to the entrainment rates of ambient air being mixed with the plume being somewhat different for different plume sizes.

To calculate the final rise for F less than 55, eq. 6, p 103 of Briggs (1971),

$$\Delta H = 21.425 \, F^{3/4} / \, u_h \qquad (3.4)$$

where u_h is the wind speed at stack top.

For F equal to or greater than 55, eq 7, p103 of Briggs (1971),

$$\Delta H = 38.71 \, F^{3/5} / u_h \qquad (3.5)$$

3.3.2 Unstable-Neutral Momentum Rise

The ΔH calculated by the appropriate above equation is then compared with the unstable-neutral momentum rise, which is calculated from eq. 5.2, p 59 from Briggs (1969):

$$\Delta H = 3 \, d \, v / u_h \qquad (3.6)$$

The highest of the momentum or buoyant rise is then used. This completes the calculation of final rise for neutral or unstable conditions.

3.4 Final Rise for Stable Conditions

3.4.1 Stable Buoyant Rise

To calculate the final plume rise for stable conditions, an intermediate variable, the stability parameter, s, must be evaluated. From p 1031 of Briggs (1971), it is:

$$s = (g \, d\theta/dz)/T \tag{3.7}$$

where
g acceleration of gravity, 9.8 m s^{-2}
$d\theta/dz$ change of potential temperature with height, K/m
T ambient air temperature, K

The change of potential temperature with height is related to the change of temperature with height by:

$$d\theta/dz = dT/dz + \Gamma \tag{3.8}$$

where Γ is the adiabatic lapse rate and equal to 0.0098 K/m. Note that the stability parameter, s, is determined only by parameter values related to the atmosphere and has no dependence upon stack conditions.

To determine the final rise under stable conditions the buoyant rise is determined from eq. 59, p 96 in Briggs(1975):

$$\Delta H = 2.6 \, [(F/(u_h \, s)]^{1/3} \tag{3.9}$$

The stable buoyancy rise for calm conditions is determined from pp 81-82 in Briggs(1975):

$$\Delta H = 4 \, F^{1/4} \, s^{-3/8} \tag{3.10}$$

The lowest of these two values is saved to compare with the stable momentum rise.

3.4.2 Stable Momentum Rise

The stable momentum rise is given by eq. 4.28, p 59 in Briggs (1969):

$$\Delta H = 1.5 \, [(v^2 \, d^2 \, T)/(4 \, T_s \, u_h)]^{1/3} \, s^{-1/6} \tag{3.11}$$

The equation for unstable-neutral momentum rise:

$$\Delta H = 3 \, d \, v/u_h \tag{3.12}$$

is also evaluated and the lower value from these latter two equations represents the final stable momentum rise.

This stable momentum rise and the stable buoyancy rise determined above are compared and the higher value taken as the final stable rise.

3.5 Final Effective Height

The effective height of the plume is determined by adding the physical height of the stack adjusted for stack-tip downwash, h', and the ΔH determined above.

3.6 Gradual Rise

As long as the buoyant plume has a temperature excess over the surrounding atmosphere, the plume will continue to rise. The period required to dissipate the excess temperature will vary with both the nature of the plume release and with atmospheric conditions, but generally will require at least three or four minutes. As a contrast to this, a momentum-dominated plume will have the momentum effect dissipated over a much smaller time period, perhaps as small as 30 to 40 seconds. Because of the short duration of the momentum rise, the momentum-dominated plume is assumed to have achieved its final rise very near the source and gradual rise is seldom considered.

However, for buoyant plumes, the gradual or transitional rise while the temperature dissipation is occurring, is estimated for all atmospheric conditions, unstable, neutral, or stable conditions, with the following single equation, which is the equivalent of eq. 2, p 1030 in Briggs (1972):

$$\Delta H = (1.60\ F^{1/3}\ x^{2/3})/u_h \tag{3.13}$$

where x is the downwind distance from the source in meters. If the rise is calculated to be more than the rise from the appropriate equation above for final rise, then it is being applied for a distance beyond the distance to final rise and the final rise value should be substituted.

3.7 Distance to Final Rise

The above equation, 3.13, can be set equal to the various equations for the estimation of final buoyant rise and solved for the distance to final rise in meters. The results are:

For unstable-neutral, for F < 55:

$$x_f = 49\ F^{5/8} \tag{3.14}$$

For unstable-neutral, for F $\geq$ 55:

$$x_f = 119F^{2/5} \tag{3.15}$$

For stable:

$$x_f = 2.07\ u\ s^{-1/2} \tag{3.16}$$

3.8 Plume Rise Used in TUPOS

In formulating the TUPOS model for elevated buoyant releases, Turner (1985) modified Briggs (1984) equations so that plume rise could be calculated layer-by-layer so that the effects of changing vertical structure of wind and temperature could be included. The methods employed are summarized in the Appendix to Turner et al. (1991).

Assuming a top-hat distribution of the plume, that is, a uniform concentration distribution from plume bottom to plume top, with the plume thickness equal to the plume rise, the top of the plume is 1.5 times the plume rise above the stack top, and the plume bottom is 0.5 times the plume rise above the stack top.

An initial estimate of the plume top, t, is determined using 1.5 times eq. 8.98, p. 351 in Briggs (1984):

$$t \quad = \quad 1.5\{1.54[F_o/(u\,u*^2)]^{2/3}\,h^{1/3}\} \tag{3.17}$$

where
F_o initial buoyancy flux at stack top, $m^4\,s^{-3}$
u wind speed at stack top, $m\,s^{-1}$
$u*$ surface friction velocity, $m\,s^{-1}$
h stack top, m

Using this estimate of the top of the plume above the stack top, an iterative estimate is made using 1.5 times eq. 8.97, p. 350 in Briggs (1984):

$$t = 1.5\{1.2[F_o/(u\,u*^2)]^{3/5}\,(h + 0.67\,t)^{2/5}\} \tag{3.18}$$

If the convective scaling parameter, $H*$, (Turner, 1986) is positive, an additional estimate is made from 1.5 times eq. 8.101, p. 351 in Briggs (1984):

$$t = 1.5\,\{3[F_o/u]^{3/5}\,H*^{-2/5}\} \tag{3.19}$$

where F_o and u are as before and $H*$ ($m^2\,s^{-3}$), the convective scaling parameter is:

$$H* \quad = \quad g\,H_s/(c_p\,\rho_a\,T_a)$$

where
g acceleration of gravity, $9.8\ m\,s^{-2}$
H_s surface sensible heat flux, watts m^{-2}
c_p specific heat at constant pressure, $0.24\ cal\,g^{-1}\,K^{-1}$
ρ_a air density, $g\,m^{-3}$
T_a air temperature, K

Using the gas law relations for density, this can be rewritten as:

$$H* \quad = \quad 0.0280\ H_s/p$$

where p atmospheric pressure in mb, and

the constant 0.0280 has units of $m^4\ mb\ watt^{-1}\ s^{-3}$

For unstable conditions the lower of these two estimates is used for the plume top. For stable conditions where $d\theta/dz$ exceeds 0.001 K m^{-1}, the lower of the two estimates above are set aside and two further equations are evaluated. The first is 1.5 times eq. 8.71, p. 344 in Briggs (1984):

$$t = 1.5 \ \{2.6[F_o/(u \ s)]^{1/3}\} \tag{3.20}$$

where
$$s = 9.806 \ (d\theta/dz)/T \tag{3.7}$$

The second is eq. 8.67, p 343 in Briggs (1984), with the coefficient set at 5.9 for estimating the plume top:

$$t = 5.9 \ [F_o/s^{3/2}]^{1/4} - 3 \ d \tag{3.21}$$

where d is the stack diameter. The lower of these two estimates is then compared with the estimate set aside above and the lower value taken as the estimate of the stable plume top.

If the estimate of the plume top is within the layer of meteorological conditions being considered, the plume rise is assumed to have been found and the plume top is t meters above the stack and the effective plume height is at 2/3 t.

If t is above this layer , it is assumed to reach into the next layer. An estimate is made of the residual buoyancy flux, F_R, existing at the top of this layer and is used for estimating the additional rise in the next layer. The F_R is estimated from one of the following four equations depending upon which equation above produced the plume rise estimate. The four equations are as follows, where z_t is the distance of the top of the layer being considered above stack top:

$$F_R = [(t - z_t)^{5/3} \ u^{*2} \ u]/[2.66 \ H^{2/3}], \tag{3.22}$$

where
$$H = h + (0.67 \ t) \tag{3.23}$$

$$F_R = [(t - z_t)^{5/3} \ H^{*2/3} \ u]/12.27 \tag{3.24}$$

$$F_R = [(t - z_t)^3 \ u \ s]/59.3 \tag{3.25}$$

$$F_R = \{[(t - z_t) + 3 \ d]^{2/5} \ s^{3/2}\}/1211.7 \tag{3.26}$$

The plume rise estimation then proceeds in the next layer solving eq. (1) - (4) with the residual buoyancy flux, F_R, substituted for F_o and using the wind speed and temperature for that layer. To obtain the new plume top estimate, the height of the bottom of this new layer above stack top is added to the result of each equation.

This is continued layer by layer until the plume top estimate is contained within a layer.

CHAPTER 4.

SPECIAL TOPICS

4.1 Conversion of Concentrations

The concentrations from dispersion equations are generally expressed in terms of mass per volume, usually grams per cubic meter. Some of the air quality sampling equipment measures air pollution in units of volume per volume. In order to compare dispersion estimates with these air quality measurements, it is necessary to convert to the same units. The basic gas law relating mass and volume through considerations of temperature and pressure is:

$$p V = m R T/M \tag{4.1}$$

where
- p atmospheric pressure, mb
- V volume, m^3
- m mass of the gas, g
- R universal gas constant, 0.0831 mb m^3 $(g\text{-mole})^{-1}$ K^{-1}
- T absolute temperature, K
- M molecular weight, g $(g\text{-mole})^{-1}$

The universal gas constant is in units so that convenient units can be used for the variables. To convert a mass, m, to a volume, the above equation can be rearranged to:

$$V = m R T/(p M) \tag{4.2}$$

So that if m is the mass in grams of a pollutant in a cubic meter of air, then V is the volume in cubic meters in a cubic meter of air. This can be rewritten as:

$$\chi \text{ (parts per part)} = \chi \text{ (g/m}^3\text{)} R T/(p M) \tag{4.3}$$

4.2 Dispersion Using Fluctuation Statistics

Following the advice of Pasquill (1961) to estimate dispersion directly from horizontal and vertical fluctuation statistics, measured horizontal standard deviations, σ_a, and vertical standard deviations, σ_e, can be used to estimate σ_y and σ_z. Hay and Pasquill (1959) and Cramer (1959, 1976) suggested ways in which this can be accomplished.

Draxler (1976) used equations with the fluctuations in radians using the small angle approximation. These can be rewritten without the small angle approximation and the fluctuation statistics expressed in degrees:

$$\sigma_y = x \tan(\sigma_a) f_y \tag{4.4}$$

$$\sigma_z = x \tan(\sigma_e) f_z \qquad\qquad (4.5)$$

where σ_a the standard deviation of the azimuth angle (wind direction) of the wind over a suitable time period, say an hour, degrees.

σ_e the standard deviation of the elevation angle of the wind over the same time period, as above, degrees.

Draxler analyzed dispersion data from a number of field experiments in order to determine the appropriate form for f_y and f_z in order to take into account stability and release height effects. Irwin (1983) also analyzed field data and suggests the form:

$$f = 1/[(1 + 0.9 \, (T/T_o)^{1/2}] \qquad\qquad (4.6)$$

for both f_y and f_z. T is travel time from source to receptor, x/u, and T_o is a time constant. T_o is 1000 for f_y; T_o is 500 for f_z for unstable (including daytime neutral) conditions; and T_o is 50 for f_z for stable (including nighttime neutral) conditions.

Note that in using these equations the dispersion is a function of wind speed as well as the turbulent fluctuations.

Since both the horizontal and vertical fluctuations vary with height above ground, it is necessary to know at what height these fluctuations should be measured. Irwin (1983) and Gryning et al. (1987) suggest that these should represent the level of the effective height of release. Turner et al. (1991) in the use of the EPRI (Electric Power Research Institute) data collected at the Kincaid power plant in central Illinois for evaluation of the model TUPOS, found that using the fluctuation statistics that are representative for half-way between the ground and the effective height of release gave better results. Since effective plume height will vary hour-by-hour in response to changes in wind speed, it is not straightforward how to place instruments to make the proper measurements. However, this can, in part, be avoided by making fluctuation measurements at set heights and by examining the vertical structure of the atmosphere, largely through similarity considerations, the fluctuation statistics for any height above ground can be inferred.

4.3 Concentrations in an Inversion Break-Up Fumigation

A surface-based inversion may be eliminated by the upward transfer of sensible heat from the ground surface when that surface is warmer than the overlying air. This situation occurs when the ground is being warmed by solar radiation or when air flows from a cold to a relatively warm surface. In either situation pollutants previously emitted above the surface into the stable layer will be mixed vertically when they are reached by the thermal eddies, and groundlevel concentrations can increase. This process, called "fumigation," was described by Hewson and Gill (1944) and Hewson (1945). Equations for estimating concentrations with these conditions have been given by Holland (1953), Hewson (1955), Gifford (1960a), Bierly and Hewson (1962), and Pooler (1965).

To estimate groundlevel concentrations under inversion break-up fumigations, one assumes that the plume was initially emitted into a stable layer. Therefore σ_y and σ_z characteristic of stable conditions must be selected for the particular distance of concern. An equation for the groundlevel concentration when the inversion has been eliminated to a height h_i is:

$$\chi_F(x,y,0;H) = \frac{Q\left[\int_{-\infty}^{p}(2\pi)^{-1/2}\exp(-0.5\,p^2)\,dp\right]}{(2\pi)^{1/2}\,\sigma_{yF}\,u\,h_i}\exp\left[-\frac{y^2}{2\,\sigma_{yF}^2}\right] \quad (4.7)$$

where $p = (h_i - H)/\sigma_z$

and σ_{yF} is discussed below.

Values for the integral in brackets can be found in most statistical tables (see Table 4.1). This factor accounts for the portion of the plume that is mixed downwind. If the inversion is eliminated up to the effective stack height, half of the plume is presumed to be mixed downward, the other half remaining in the stable layer above. Eq. 4.7 can be approximated when the fumigation concentration is near its maximum by:

$$\chi_F(x,y,0;H) = \frac{Q}{(2\pi)^{1/2}\,\sigma_{yF}\,u\,h_i}\exp\left[-\frac{y^2}{2\,\sigma_{yF}^2}\right] \quad (4.8)$$

where $h_i = H + 2\,\sigma_z = h + \Delta H + 2\,\sigma_z$

A difficulty is encountered in estimating a reasonable value for the horizontal dispersion, because in mixing the stable plume through a vertical depth some additional horizontal spreading occurs (see Problem 23 in Chapter 8). If this spreading is ignored and the σ_y for stable conditions used, the probable result would be estimated concentrations higher than actual concentrations. Or, using an approximation suggested by Bierly and Hewson (1962) that the edge of the plume spreads outward with an angle of 15°, the σ_{yF} for the inversion break-up fumigation equals the σ_y for stable conditions plus one-eighth the effective height of emission. The origin of this concept can be seen in Figure 4.1 and eq. 4.9, where the edge of the plume is the point at which the concentration falls to 1/10 that at the centerline (at a distance of 2.15 σ_y from the plume center).

$$\sigma_{yF} = \frac{2.15\,\sigma_y(\text{stable}) + H\tan 15°}{2.15} \quad (4.9)$$

$$= \sigma_y(\text{stable}) + H/8$$

A Gaussian distribution in the horizontal is assumed.

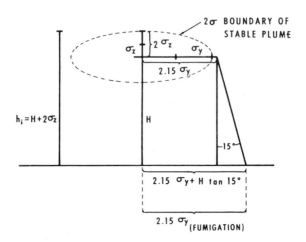

Figure 4.1 Diagram showing assumed height, h_i, and σ_y during fumigation, for use in Equation 4.9.

Eq. 4.8 should not be applied near the stack, for if the inversion has been eliminated to a height sufficient to include the entire plume, the emission is taking place under unstable, not stable, conditions. Therefore, the nearest downwind distance to be considered for an estimate of fumigation concentrations must be great enough, based on the time required to eliminate the inversion, that this portion of the plume was initially emitted into stable air. This distance is $x = u\, t_m$, where u is the mean wind in the stable layer and t_m is the time required to eliminate the inversion from h, the physical height of the stack to h_i.

t_m is dependent upon both the strength of the inversion and the rate of heating at the surface. Pooler (1965) has derived an expression for estimating this time:

$$t_m = \frac{\rho_a\, c_p}{R} \; \frac{d\theta}{dz} \; (h_i - h)\, [(h + h_i)/2] \qquad (4.10)$$

where

t_m time required for the mixing layer to develop from the top of the stack to the top of the plume, sec

ρ_a ambient air density, g m^{-3}

c_p specific heat of air at constant pressure, cal g^{-1} K^{-1}

R   net rate of sensible heating of an air column by solar radiation, cal m^{-2} s^{-1}

$d\theta/dz$ vertical potential temperature gradient, $K\ m^{-1}$, approximately equal to $dT/dz + \Gamma$ where Γ is the adiabatic lapse rate, approximately $0.0098\ K\ m^{-1}$

h_i height of base of the inversion sufficient to be above the plume, m

h physical height of the stack, m

Note that $h_i - h$ is the thickness of the layer to be heated and $(h + h_i)/2$ is the average height of the layer. Although R depends on season and cloud cover, and varies continuously with time, Pooler (1965) has used a value of 67 cal $m^{-2}\ s^{-1}$ as an average for fumigation.

Hewson (1945) also suggested a method of estimating the time required to eliminate an inversion to a height z by use of an equation of Taylor's (p 8, 1915):

$$t = z^2/(4\ K) \tag{4.11}$$

where   t   time required to eliminate the inversion to height z, s

z   height to which the inversion has been eliminated, m

K   eddy diffusivity for heat, $m^2\ s^{-1}$

Rewriting to compare with (4.10),

$$t_m \;=\; (h_i^2 - h^2)/(4\ K) \tag{4.12}$$

Hewson has suggested a value of 3 $m^2\ s^{-1}$ for K.

4.4 Partial Penetration of Plumes Through the Mixing Height

If the mixing height is not too high and there is sufficient buoyancy in the plume, part of the plume may penetrate into the inversion above the mixing height. Briggs (1975) suggested ways of approximating the amount of the plume that penetrates. This procedure assumes that the concentration distribution vertically is uniform from bottom to top, a "top hat" distribution, and that the thickness of the plume is equal to the plume rise.

Several different procedures for the calculation of partial penetration are discussed by Weil (1988, p. 135- 147).

Using the plume rise through layers given in Chapter 3, the plume top and bottom can be determined. The mixing height minus the plume bottom divided by the plume top minus the plume bottom will give the portion of the plume beneath the mixing height.

Table 4.1 Areas for the cumulative normal distribution, that is, from the left side (- infinity).

s	0.0	0.1	0.2	0.3	0.4	0.5	0.6	0.7	0.8	0.9
-4.	.00003	.00002	.00001	.00001	.00001	.00000	.00000	.00000	.00000	.00000
-3.	.00135	.00097	.00069	.00048	.00034	.00023	.00016	.00011	.00007	.00005
-2.	.02275	.01787	.01390	.01072	.00820	.00621	.00466	.00347	.00256	.00187
-1.	.15866	.13567	.11507	.09680	.08076	.06681	.05480	.04457	.03593	.02872
-0.	.50000	.46017	.42074	.38209	.34458	.30854	.27425	.24197	.21186	.18406
+0.	.50000	.53983	.57926	.61791	.65542	.69146	.72575	.75803	.78814	.81594
+1.	.84134	.86433	.88493	.90320	.91924	.93319	.94520	.95543	.96407	.97128
+2.	.97725	.98213	.98610	.98928	.99180	.99379	.99534	.99653	.99744	.99813
+3.	.99865	.99903	.99931	.99952	.99966	.99977	.99984	.99989	.99993	.99995
+4.	.99997	.99998	.99999	.99999	.99999	1.00000	1.00000	1.00000	1.00000	1.00000

4.5 Concentrations at Groundlevel Compared to Concentrations at the Level of Effective Stack Height from Elevated Continuous Sources

There are several interesting relationships between groundlevel concentrations and concentrations at the level of the plume centerline. One of these is at the distance of maximum concentration at the ground. As a rough approximation the maximum groundlevel concentration occurs at the distance where $\sigma_z = (1/2^{1/2})$ H. This approximation is much better for unstable conditions than for stable conditions. With this approximation, the ratio of concentration at plume centerline to that at the ground is:

$$\frac{\chi(x,0,H)}{\chi(x,0,0)} = \frac{0.5\,\{1.0 + \exp\,[-0.5(2H/\sigma_z)^2]\}}{\exp\,[-0.5(H/\sigma_z)^2]} \tag{4.13}$$

Substituting for the ratio of H to σ_z of $2^{1/2}$, the above equals 1.38.

This calculation indicates that at the distance of maximum groundlevel concentration the concentration at plume centerline is greater by about one-third.

It is also of interest to determine the relationship between σ_z and H such that the concentration at groundlevel at a given distance from the source is about the same as the concentration at plume level. This condition should occur where:

$$\exp\left[-\frac{H^2}{2\,\sigma_z^2}\right] = 0.5\left\{1 + \exp\left[-\frac{(2H)^2}{2\,\sigma_z^2}\right]\right\} \tag{4.14}$$

The value $H/\sigma_z = 1.10$ satisfies this expression, which can be written as $\sigma_z = 0.91$ H (see Problem 21 in Chapter 8).

4.6 Total Dosage from a Finite Release

The total dosage, which is the integration of concentration over the time of passage of a plume or puff, can be obtained from:

$$D_T(x,y,0;H) = \frac{Q_T}{\pi\,u\,\sigma_y\,\sigma_z}\exp\left[-\frac{y^2}{2\,\sigma_y^2}\right]\exp\left[-\frac{H^2}{2\,\sigma_z^2}\right] \tag{4.15}$$

where $\quad D_T \quad = \quad$ total dosage, $g\ s\ m^{-3}$
and $\quad\ \ Q_T \quad = \quad$ total release, g

The σ's should be representative of the time period over which the release takes place, and care should be taken to consider the x-axis along the trajectory or path of the plume or puff travel. Large errors can easily occur if the path is not known accurately. The estimate of this path is usually increasingly difficult with shorter release times. D_T can also be given in curie s m^{-3} if Q_T is in curies.

4.7 Crosswind-Integrated Concentration

The groundlevel crosswind-integrated concentration is often of interest. For a continuous elevated source this concentration is determined from Equation 2.3 integrated with respect to y from -∞ to +∞ (Gifford, 1960a) giving:

$$\chi_{CWI} = \frac{2\ Q}{(2\pi)^{0.5}\ u\ \sigma_z}\ \exp\left[-\frac{H^2}{2\ \sigma_z^2}\right] \qquad (4.16)$$

In diffusion experiments the groundlevel crosswind-integrated concentration is often determined at particular downwind distances from a crosswind line or arc of sampling measurements made at this distance. When the source strength, Q, and average wind speed, u, are known, σ_z can be estimated indirectly even though no measurements were made in the vertical. If any of the tracer is lost through reaction or deposition, the resulting σ_z from such estimates will not represent the proper vertical dispersion (see Problem 29).

4.8 Variation of Concentrations with Sampling Time

Concentrations directly downwind from a source decrease with sampling time mainly because of a larger σ_y due to increased meander of wind direction. Stewart, Gale, and Crooks (1958) reported that this decrease in concentration follows a one-fifth power law with the sampling time for sampling periods from about 3 minutes to about half an hour. Cramer (1959) indicates that this same power law applies for sampling times from 3 seconds to 10 minutes. Both of these studies were based on observations taken near the height of release. Gifford (1960b) indicates that ratios of peak to mean concentrations are much higher than those given by the above power law where observations of concentrations are made at heights considerably different from the height of release or considerably removed from the plume axis. He also indicates that for increasing distances from an elevated source, the ratios of peak to average concentrations observed at groundlevel approach unity. Singer (1961) and Singer et al. (1963) show that ratios of peak to mean concentrations depend also on the stability of the atmosphere and the type of terrain that the plume is passing over. Nonhebel (1960) reports that Meade deduced a relation between calculated concentrations at groundlevel and the sampling time from "a study of published data on lateral and vertical diffusion coefficients in steady winds." These relations are shown in Table 4.2.

This table indicates a power relation with time: $\chi \propto t^{-0.17}$. Note that these estimates were based upon published dispersion coefficients rather than upon sampling results. Information in the references cited indicates that effects of sampling time are exceedingly

Table 4.2. Variation of Calculated Concentration with Sampling Time

Sampling Time	Ratio of Calculated Concentration to Three-Minute Concentration
3 minutes	1.00
15 minutes	0.82
1 hour	0.61
3 hours	0.51
24 hours	0.36

complex. If it is necessary to estimate concentrations from a single source for time intervals greater than a few minutes, the best estimate apparently can be obtained from:

$$\chi_s = \chi_k (t_k/t_s)^p \tag{4.17}$$

where χ_s desired concentration estimate for the sampling time, t_s

χ_k concentration estimated for the shorter sampling time, t_k (probably about 3 minutes)

p   between 0.17 and 0.2

Eq. 4.17 probably would be applied most appropriately to sampling times less than two hours.

Hino (1968) examined the relationships between groundlevel concentrations and sampling times. He found that for the sampling times between 10 minutes and five to six hours the concentrations were proportional to sampling time to the -0.5 power. He also indicates that the -0.2 power law reported by Nonhebel (1960) is most likely valid for sampling times less than 10 minutes. Table 4.3 results from values read from Hino's Figure 9, show the ratios of concentrations for various averaging times to that for one hour (see Problem 30).

Table 4.3. Peak to Mean Concentration Ratios from Hino (1968)

Period	Peak to One-Hour
One hour	1.0
30 Minutes	1.3
10 Minutes	2.3
3 Minutes	4.
1 Minute	4. to 7.
30 Seconds	4. to 10.

4.9 Estimation of Seasonal or Annual Concentrations at a Receptor from a Single Pollutant Source

For a source that emits at a constant rate from hour to hour and day to day, estimates of seasonal or annual average concentrations can be made for any distance in any direction if "stability wind rose" data (commonly referred to as STAR, STability ARray, data) are available for the period under study. A wind rose gives the frequency of occurrence for each wind direction (usually to 16 points) and wind speed class (frequently in six classes) for the period under consideration (from 3 months to 10 years). A stability wind rose gives the same type of information for each Pasquill stability class.

If the wind directions are specified to 16 points and it is assumed that the wind directions within each sector are distributed randomly over a period of a month or a season, it can further be assumed that the effluent is uniformly distributed in the horizontal within the corresponding downwind sector (p 540, Holland, 1953). The appropriate equation for average concentration for a specific wind speed and stability is then either :

$$\bar{\chi} = \frac{2\,Q}{(2\pi)^{1/2}\,\sigma_z\,u\,[(2\pi\,x)/16]}\;\exp\left[-\frac{H^2}{2\,\sigma_z^{\,2}}\right] \tag{4.18}$$

$$= \frac{2.03\,Q}{\sigma_z\,u\,x}\;\exp[-0.5(H/\sigma_z)^2]$$

or

$$\chi = \frac{Q}{z_i\,u\,[(2\pi\,x)/16]} = \frac{2.55\,Q}{z_i\,u\,x} \tag{4.19}$$

depending upon whether a stable layer aloft is affecting the distribution.

The estimation of χ for a particular direction and downwind distance can be accomplished by choosing a representative wind speed for each speed class and solving the appropriate equation for all wind speed classes and stabilities. Note that a SSW wind affects receptors to the NNE of a source. One obtains the average concentration for a given direction and distance by determining all the concentrations and weighting each one according to its frequency for the particular stability and wind speed class and summing. If desired, a different effective height of emission can be used for various wind speeds. The average concentration can be expressed by:

$$\chi(x,\theta) = \sum_S \sum_N \frac{2\,Q\,f(\theta,S,N)}{(2\pi)^{1/2}\,\sigma_{zS}\,u_N\,[(2\pi\,x)/16]}\;\exp\left[-\frac{H_u^{\,2}}{2\,\sigma_{zS}^{\,2}}\right] \tag{4.20}$$

where $f(\theta,S,N)$ frequency during the period of interest that the wind is from the direction, θ, for the stability condition, S, and the wind speed class, N

σ_{zS} vertical dispersion parameter evaluated at the distance x for the stability condition S

H_u effective height of release for the wind speed, u_N

Where stability wind rose information cannot be obtained, a first-order approximation may be made of seasonal or annual average concentrations by using the appropriate wind rose in the same manner, and assuming the neutral stability class, D, only.

4.10 Meteorological Conditions Associated with Maximum Concentrations

For groundlevel sources the maximum concentrations occur with stable conditions.

For elevated sources, maximum *instantaneous* concentrations occur with unstable conditions when portions of the plume that have undergone little dispersion are brought to the ground. These occur close to the point of emission (on the order of 1 to 3 stack heights). These concentrations are usually of little general interest because of their very short duration; they *cannot* be estimated from the material presented in this workbook.

For elevated sources, maximum concentrations *for time periods of a few minutes* occur with unstable conditions; although the concentrations fluctuate considerably under these conditions, the concentrations averaged over a few minutes are still high compared to those found under other conditions. The distance of this maximum concentration occurs near the stack (from 1 to 5 effective heights of emission downwind) and the concentration drops off rapidly downwind with increasing distance.

For elevated sources, maximum concentrations *for time periods of about half an hour* can occur with fumigation conditions when an unstable layer increases vertically due to surface heating to mix downward a plume previously discharged within a stable layer. With small plume rise, ΔH, the fumigation can occur close to the source but will be of relatively short duration. For large plume rise, the fumigation will occur some distance from the stack (perhaps tens of kilometers) but can persist for a longer time interval. Concentrations considerably lower than those associated with fumigations, but of significance, can occur with neutral or unstable conditions when the dispersion upward is severely limited by the existence of a more stable layer above the plume, for example, an inversion. This is usually referred to as plume trapping.

Under stable conditions the maximum concentrations at groundlevel from elevated sources are less than those occurring under unstable conditions and occur at greater distances from the source. However, the difference between maximum groundlevel concentrations for stable and unstable conditions is only a factor of 2 for effective heights of 25 meters and a factor of 5 for H of 75 m. Because the stable maximum occurs at greater distances, concentrations that are below the maximum but still significant can occur over large areas. This becomes increasingly significant if emissions are coming from more than one source.

4.11 Dual Effect of Wind Speed —
Maximum Concentrations at the Critical Wind Speed

Wind speed has two effects on a source that has buoyancy or momentum rise: dilution and plume rise. If there is no plume rise the highest groundlevel concentrations will occur with the lowest wind speed when the dilution is least. But if there is momentum or buoyant rise, the light wind conditions will, in addition to the minimum dilution, also cause the greatest plume rise. At contrast to this is the high wind situation, which will cause the lowest plume rise but will cause a large dilution. Between these extremes is a wind speed that will cause an intermediate plume rise and an intermediate dilution and will result in the maximum concentration. This wind speed that causes the highest concentration from this source for a given stability condition is called the critical wind speed. The computer code, PTPLU (Pierce et al., 1982), was designed to make calculations for the different possible wind speed-Pasquill stability class combinations. From inspection of the results from this model, it is possible to approximate the critical wind speed for each stability.

4.12 Concentrations at a Receptor Point from Several Sources

Sometimes, especially for multiple sources, it is convenient to consider the receptor as being at the origin of the dispersion coordinate system. The source receptor geometry can then be worked out merely by drawing or visualizing an x-axis oriented upwind from the receptor and determining the crosswind distance of each source in relation to this x-axis. As pointed out by Gifford (1959), the concentration at (0,0,0) from a source at (x,y,H) on a coordinate system with the x-axis oriented upwind is the same as the concentration at (x,y,0) from a source at (0,0,H) on a coordinate system with the x-axis downwind (Figure 4.2). The total concentration is then given by summing the individual contributions from each source (see problem 31).

It is often difficult to determine the atmospheric conditions of wind direction, wind speed, and stability that will result in the maximum combined concentrations from two or more sources. This is why most regulatory approved models determine such concentrations as the second highest that occurs with a frequency of once-per-year by making the simulations for each hour of the year and determining second highest from the calculations of each concentration for a given averaging time and saving the highest and second highest as the calculations are made.

4.13 Rural and Urban Dispersion Parameters

The Pasquill-Gifford dispersion parameters which were introduced in Chapter 2 are considered to be appropriate for rural areas. Pasquill (1976) indicated that these were valid for a surface roughness length, z_0, of 3 cm.

McElroy and Pooler (1968) analyzed the series of tracer experiments conducted in St. Louis and reported the results. Considering their data and results, Briggs (1973) suggested the equations given in Table 4.4 for urban conditions. These are shown in

Figure 4.3 from Gifford (1976). Values for various source to receptor distances are given in Table 4.5 at the end of the chapter. For most atmospheric conditions these dispersion parameters for urban conditions are larger than those for rural conditions, reflecting the influences of increased mechanical turbulence in the urban area and buoyant turbulence during the evening and nighttime hours as the result of the released urban heat stored in the structures and pavement during the daytime. The appropriate curve is selected on the basis of a Pasquill stability class determined in the rural environment near the urban area.

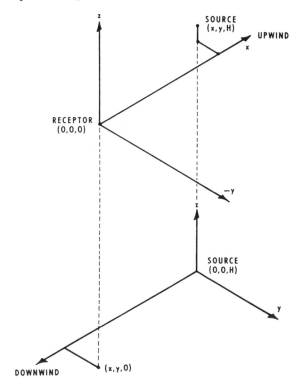

Figure 4.2 Comparison of source-oriented and receptor-oriented coordinate systems.

4.14 Exponent for Power Law Wind Profiles

Specific exponents are frequently used in the power-law equation (equation 1.1) for increase of wind speed with height corresponding to each Pasquill stability class. These are given in Table 4.6 for rural and urban conditions.

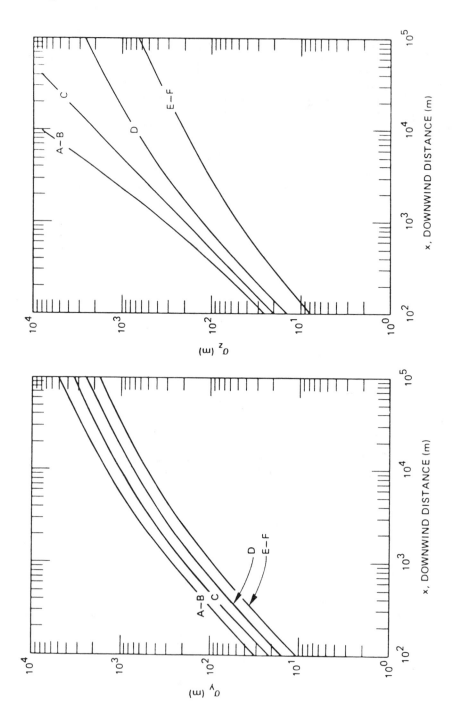

Figure 4.3 Briggs urban dispersion parameters, σ_y and σ_z. Source: Gifford (1976)

Table 4.4 Urban Dispersion Parameters by Briggs.

Pasquill type	σ_y m.	σ_z m.
A-B	$0.32 \, x \, (1 + 0.0004 \, x)^{-0.5}$	$0.24 \, x \, (1 + 0.001 \, x)^{0.5}$
C	$0.22 \, x \, (1 + 0.0004 \, x)^{-0.5}$	$0.20 \, x$
D	$0.16 \, x \, (1 + 0.0004 \, x)^{-0.5}$	$0.14 \, x \, (1 + 0.0003 \, x)^{-0.5}$
E-F	$0.11 \, x \, (1 + 0.0004 \, x)^{-0.5}$	$0.08 \, x \, (1 + 0.0015 \, x)^{-0.5}$

For distances, x, between 100 and 10000 meters.

Table 4.6 Power-Law Wind Profile Exponents.

Pasquill Stability Class	Rural Wind Profile Exponent	Urban Wind Profile Exponent
A	0.07	0.15
B	0.07	0.15
C	0.10	0.20
D	0.15	0.25
E	0.35	0.30
F	0.55	0.30

4.15 Buoyancy-Induced Dispersion

The Gaussian point-source equations as used with the Pasquill-Gifford parameters for rural conditions or the Briggs-urban parameters treat the source as if it was a point source of very small size. The buoyancy-induced dispersion provides a method for accounting for the finite size of the source. The principal effect that causes the rapid growth of the plume upon release is the entrainment of ambient air into the plume.

If the plume is dominated by momentum the shearing motion between the vertical rising plume and the outside air causes the entrainment. Buoyantly rising plumes have a boiling type of action with circular eddies entraining outside air into the rising plume. For the inclusion of these effects, the initial size of the plume is scaled according to the gradual plume rise. This provides a small effect near the source and the maximum effect at the point of maximum rise. As the plume is affected by the wind and becomes horizontal it is assumed to be basically symmetrical in the horizontal and the vertical as the plume reaches final rise.

Therefore, these effects near the point of release are assumed to cause initial horizontal and vertical plume sizes that are equal. The equation for these initial sizes is:

$$\sigma_{yo} \ = \ \sigma_{zo} \ = \ \Delta H/3.5 \qquad\qquad (4.21)$$

The ΔH used is that for gradual rise for downwind distances less than the distance to final rise. The final ΔH is used for distances at and beyond the distance to final plume rise. The σ_{yo} and σ_{zo} are used with the dispersion parameters in the following way to result in the effective dispersion:

$$\sigma_{ye} \ = \ [\sigma_y{}^2 \ + \ \sigma_{yo}{}^2]^{0.5} \qquad\qquad (4.22)$$

$$\sigma_{ze} \ = \ [\sigma_z{}^2 \ + \ \sigma_{zo}{}^2]^{0.5} \qquad\qquad (4.23)$$

The effect of including buoyancy-induced dispersion is to increase the dispersion decreasing the concentrations in the plume. This effect is, of course, more noticeable near the source, where the dispersion due to ambient turbulence is small. For most sources the effect on groundlevel concentrations is small. For calculations of concentration above the ground, especially near plume level, the calculated concentrations will be more realistic with the inclusion of this additional plume spreading.

4.16 Area Sources

In determining concentrations resulting from a large number of sources, such as in urban areas, there may be too many sources to consider the dispersion from each one individually. Often an approximation can be made by combining all of the emissions in a given area and treating this area as a source having an initial horizontal standard deviation, σ_{yo}. A virtual distance, x_y, can then be found that will give this horizontal standard deviation. This is just the distance that will yield the appropriate value for σ_y for this stability. Then equations for point sources may be used, determining σ_y as a function of $x + x_y$, a slight variation of the suggestion by Holland (1953). This procedure treats the area source as a crosswind line source with a normal distribution, a fairly good approximation for the distribution across an area source. The initial standard deviation for a square area source can be approximated by $\sigma_{yo} = s/4.3$, where s is the length of a side of the area (see Problem 33).

If the emissions within an area are from varying effective stack heights, the variation may be approximated by using a σ_{zo}, the standard deviation of the initial vertical distribution of sources. A virtual distance, x_z, can be found, and point source equations used for estimating concentrations, determining σ_z as a function of $x + x_z$.

4.17 Topography

Under conditions of irregular topography the direct application of a standard dispersion equation is often invalid. In some situations the best one may be able to do without the benefit of *in situ* experiments is to estimate the upper limit of the concentrations likely to occur.

For example, to calculate the concentrations on a hillside downwind from and facing the source and at the effective source height, the equation for concentrations at the centerline of the plume (eq. 2.4) will yield the highest expected concentrations. This would closely approximate the situation under stable conditions, when the pollutant plume is most likely to encounter the hillside. Under unstable conditions the flow is more likely to rise over the hill (see Problem 32).

With downslope flow when the receptor is at a lower elevation than the source, a likely assumption is that the flow parallels the slope; i.e., no allowance is made for the difference between groundlevel elevations at the source and at the receptor.

Where a steep ridge or bluff restricts the horizontal dispersion, the flow is likely to be parallel to such a bluff. An assumption of complete eddy reflection at the bluff, similar to eddy reflection at the ground from an elevated source, is in order. This may be accomplished by using:

$$\chi(x,y,0;H) = \frac{Q}{\pi\, u\, \sigma_y\, \sigma_z} \left\{ \exp\left[-\frac{y^2}{2\,\sigma_y{}^2} \right] + \exp\left[-\frac{(2B - y)^2}{2\,\sigma_y{}^2} \right] \right\}$$
$$\left\{ \exp\left[-\frac{H^2}{2\,\sigma_z{}^2} \right] \right\} \qquad (4.24)$$

B is the distance from the x-axis to the restricting bluff, and the positive y-axis is defined to be in the direction of the bluff.

The restriction of horizontal dispersion of down-valley flow by valley sides is somewhat analogous to restriction of the vertical dispersion by a stable layer aloft. When the σ_y becomes great enough, the concentrations can be assumed to be uniform across the width of the valley and the concentration calculated according to the following equation, where in this case Y is the width of the valley.

$$\bar{\chi} = \frac{2\,Q}{(2\pi)^{1/2}\, \sigma_z\, Y\, u} \exp\left[-\frac{H^2}{2\,\sigma_z{}^2} \right] \qquad (4.25)$$

4.18 Line Sources

Concentrations downwind of a continuously emitting infinite line source, when the wind direction is normal to the line, can be expressed by rewriting equation (12) p 154 of Sutton (1932):

$$\chi(x,y,0;H) = \frac{2\,q}{(2\pi)^{1/2}\,\sigma_z\,u} \exp\left[-\frac{H^2}{2\,\sigma_z^2}\right] \qquad (4.26)$$

Here q is the source strength per unit distance, for example, $g\ s^{-1}\ m^{-1}$. Note that the horizontal dispersion parameter, σ_y, does not appear in this equation, since it is assumed that lateral dispersion from one segment of the line is compensated by dispersion in the opposite direction from adjacent segments. Also, y does not appear, since concentration at a given x is the same for any value of y (see Problem 34).

Concentrations from infinite line sources when the wind is not perpendicular to the line can be approximated. If the angle between the wind direction and line source is φ, the equation for concentration downwind of the line source is:

$$\chi(x,y,0;H) = \frac{2\,q}{\sin\varphi\,(2\pi)^{1/2}\,\sigma_z\,u} \exp\left[-\frac{H^2}{2\,\sigma_z^2}\right] \qquad (4.27)$$

The horizontal dispersion parameter, σ_y, is a function of the downwind distance of the receptor from the upwind point on the line source, not the perpendicular distance. This equation should not be used where φ is less than 45°.

For situations with the angle φ less than 45°, there is a point on the line directly upwind at a distance x from the receptor. The emissions in the vicinity of this point have the major impact upon the receptor. However, there are also points on the line at closer distances that affect the receptor, whose vertical dispersion, σ_z, will be smaller as the material reaches the receptor than the material that traveled x. Also, there will be points on the line at greater distances than x that affect the receptor, whose vertical dispersion, σ_z, will be larger as the material reaches the receptor than the material that traveled x. Therefore, one σ_z will be inadequate for each of the portions of the line affecting the receptor. For wind flow more parallel to the line than 45°, a model should be used that will numerically integrate the effects of the various portions of the line source. Such models are PAL (Petersen and Rumsey, 1987), HIWAY2 (Petersen, 1980), and CALINE3 (Benson, 1979).

When estimating concentrations from finite line sources, one must account for "edge effects" caused by the ends of the line source. These effects will, of course, extend to greater crosswind distances as the distance from the source increases. For concentrations from a finite line source oriented crosswind, define the x-axis in the direction of the mean wind and passing through the receptor of interest. The limits of the line source can be defined as extending from y_1 to y_2 where y_1 is less than y_2. The equation (11), p 154 from Sutton (1932) is:

$$\chi(x,y,0;H) = \frac{2\,q}{(2\pi)^{1/2}\,\sigma_z\,u}\;\exp\left[-\frac{H^2}{2\,\sigma_z^2}\right]$$

$$\int_{p_1}^{p_2}\frac{1}{(2\pi)^{1/2}}\exp(-0.5\,p^2)\;dp \qquad (4.28)$$

where $p_1 = y_1/\sigma_y$, and $p_2 = y_2/\sigma_y$.

The value of the integral can be determined from Table 4.1. (Also see Problem 35.)

4.19 Instantaneous Sources

Thus far we have considered only sources that were emitting continuously or for time periods equal to or greater than the travel times from the source to the point of interest. Cases of instantaneous release, as from an explosion, or short-term releases on the order of seconds, are often of practical concern. To determine concentrations at any position downwind, one must consider the time interval after the time of release and diffusion in the downwind direction as well as lateral and vertical dispersion. Of considerable importance, but very difficult, is the determination of the path or trajectory of the "puff." This is most important if concentrations are to be determined at specific points. Determining the trajectory is of less importance if knowledge of the magnitude of the concentrations for particular downwind distances or travel times is required without the need to know exactly at what points these concentrations occur. Rewriting equation (13) p 155 of Sutton (1932), results in an equation that may be used for estimates of instantaneous concentration downwind from a release from height, H:

$$\chi(x,y,0;H) = \frac{2\,Q_T}{(2\pi)^{3/2}\,\sigma_x\,\sigma_y\,\sigma_z}\;\exp\left[-\frac{(x-ut)^2}{2\,\sigma_x^2}\right]$$

$$\exp\left[-\frac{y^2}{2\,\sigma_y^2}\right]\;\exp\left[-\frac{H^2}{2\,\sigma_z^2}\right] \qquad (4.29)$$

The numerical value of $(2\pi)^{3/2}$ is 15.75.

The symbols have the usual meaning, with the important exceptions that Q_T represents the *total mass* of the release and the σ's are *not* those evaluated with respect to the dispersion of a continuous source at a fixed point in space, such as the Pasquill-Gifford or Briggs Urban parameters.

The equation for the concentration as the center of the puff passes the downwind position at the distance x is:

$$\chi(x,0,0;H)_{MAX} = \frac{2\,Q_T}{(2\pi)^{3/2}\,\sigma_x\,\sigma_y\,\sigma_z}\,\exp\left[-\frac{H^2}{2\,\sigma_z^2}\right] \quad (4.30)$$

In equations 4.24 and 4.25 the σ's refer to dispersion statistics following the motion of the expanding puff. The σ_x is the standard deviation of the concentration distribution in the puff in the upwind-downwind direction, and t is the time after release. Note that there is no dilution in the downwind direction by wind speed. The speed of the wind mainly serves to give the downwind position of the center of the puff, as shown by examination of the exponential involving σ_x. Wind speed may influence the dispersion indirectly because the dispersion parameters σ_x, σ_y, and σ_z may be functions of wind speed. The σ_y's and σ_z's for an instantaneous source are less than those for a few minutes given in Figures 2.3 and 2.4. Slade (1965) has suggested values for σ_y and σ_z for quasi-instantaneous sources. These are given in Table 4.7. The problem remains to make best estimates of σ_x. Much less is known of dispersion in the upwind-downwind direction than is known of lateral and vertical dispersion. In general, one should expect the σ_x value to be about the same as σ_y. Initial dimensions of the puff, i.e., from an explosion, may be approximated by finding a virtual distance to give the appropriate initial standard deviation for each direction. Then σ_y will be determined as a function of $x + x_y$, σ_z as a function of $x + x_z$, and σ_x as a function of $x + x_x$.

Table 4.7. Dispersion Parameters for Quasi-Instantaneous Sources (From Slade, 1965)

Stability	x = 100 m		x = 4 km	
	σ_y	σ_z	σ_y	σ_z
Unstable	10	15	300	220
Neutral	4	3.8	120	50
Very Stable	1.3	0.75	35	7

If one assumes that the Unstable conditions in the above table are closely approximately by Pasquill Class B stability, Neutral by class D, and Very Stable by "G", then approximate power-law functions, $\sigma_y = a\,x^b$ and $\sigma_z = c\,x^d$, can be specified for quasi-instantaneous sources corresponding to the Pasquill stability classes. The coefficients and powers are given in Table 4.8.

Table 4.8. Quasi-Instantaneous Power Functions

Pasquill Stability	a	b	σ_y 100 m	σ_y 4 km.	c	d	σ_z 100 m	σ_z 4 km.
A	0.18	0.92	12.45	371	0.72	0.76	23.8	393
B	0.14	0.92	9.69	288	0.53	0.73	15.3	226
C	0.1	0.92	6.92	206	0.34	0.72	9.4	133
D	0.06	0.92	4.15	124	0.15	0.70	3.8	50
E	0.045	0.91	2.97	85	0.12	0.67	2.6	31
F	0.03	0.90	1.89	52	0.08	0.64	1.5	16
"G"	0.02	0.89	1.21	32	0.05	0.61	0.8	8

4.20 Average Concentrations from Instantaneous Sources

The highest average concentrations over the time period, τ, from an instantaneous puff release, that is, when the center of the puff passes the receptor at the midpoint of the period, can be estimated by making a calculation of the instantaneous concentration for this receptor at the center of the puff and applying a correction factor which is the ratio of the average to the peak concentration. The first step in calculating this correction factor is to determine half the number of standard deviations, N, that pass by the receptor during the averaging time:

$$N = (\tau u)/(2 \sigma_x) \tag{4.31}$$

where u is the wind speed and σ_x is the alongwind dispersion parameter value at the downwind distance of the receptor location. The correction factor, F, can then be determined from:

$$F = (A - 0.5)/(0.3989 N) \tag{4.32}$$

where A is the cumulative area under the normal curve to N deviations. These cumulative areas are given in Table 4.1. This procedure is from Petersen (1982). The concentration at a receptor location directly downwind from the release point begins at zero, rises to the maximum concentration as the center of the puff passes, and then drops, eventually becoming zero again.

Table 4.5 (Part 1) Briggs Urban Dispersion Parameters

x, km	Sigma-y, meters				Sigma-z, meters				Sigma-y times Sigma-z			
	A-B	C	D	E-F	A-B	C	D	E-F	A-B	C	D	E-F
0.01	3.19	2.20	1.60	1.10	2.41	2.00	1.40	0.79	7.70	4.39	2.23	0.87
0.02	6.37	4.38	3.19	2.19	4.85	4.00	2.79	1.58	30.9	17.5	8.90	3.45
0.03	9.54	6.56	4.77	3.28	7.31	6.00	4.18	2.35	69.7	39.4	20.0	7.70
0.04	12.7	8.73	6.35	4.37	9.79	8.00	5.57	3.11	124.	69.8	35.3	13.6
0.05	15.8	10.9	7.92	5.45	12.3	10.00	6.95	3.86	195.	109.	55.0	21.0
0.06	19.0	13.0	9.49	6.52	14.8	12.0	8.33	4.60	281.	157.	79.0	30.0
0.07	22.1	15.2	11.0	7.59	17.4	14.0	9.70	5.33	384.	213.	107.	40.5
0.08	25.2	17.3	12.6	8.66	20.0	16.0	11.1	6.05	503.	277.	139.	52.4
0.09	28.3	19.5	14.1	9.73	22.6	18.0	12.4	6.76	638.	350.	176.	65.7
0.10	31.4	21.6	15.7	10.8	25.2	20.0	13.8	7.46	790.	431.	216.	80.5
0.11	34.5	23.7	17.2	11.8	27.8	22.0	15.2	8.15	958.	521.	261.	96.6
0.12	37.5	25.8	18.8	12.9	30.5	24.0	16.5	8.84	1.14E+03	619.	310.	114.
0.13	40.6	27.9	20.3	13.9	33.2	26.0	17.9	9.51	1.35E+03	725.	362.	133.
0.14	43.6	30.0	21.8	15.0	35.9	28.0	19.2	10.2	1.56E+03	839.	419.	153.
0.15	46.6	32.1	23.3	16.0	38.6	30.0	20.5	10.8	1.80E+03	962.	479.	174.
0.16	49.6	34.1	24.8	17.1	41.4	32.0	21.9	11.5	2.05E+03	1.09E+03	543.	196.
0.17	52.6	36.2	26.3	18.1	44.1	34.0	23.2	12.1	2.32E+03	1.23E+03	611.	220.
0.18	55.6	38.2	27.8	19.1	46.9	36.0	24.5	12.8	2.61E+03	1.38E+03	683.	244.
0.19	58.6	40.3	29.3	20.1	49.7	38.0	25.9	13.4	2.92E+03	1.53E+03	758.	270.
0.20	61.6	42.3	30.8	21.2	52.6	40.0	27.2	14.0	3.24E+03	1.69E+03	837.	297.
0.21	64.5	44.4	32.3	22.2	55.4	42.0	28.5	14.7	3.58E+03	1.86E+03	920.	325.
0.22	67.5	46.4	33.7	23.2	58.3	44.0	29.8	15.3	3.94E+03	2.04E+03	1.01E+03	354.
0.23	70.4	48.4	35.2	24.2	61.2	46.0	31.1	15.9	4.31E+03	2.23E+03	1.10E+03	384.
0.24	73.4	50.4	36.7	25.2	64.1	48.0	32.5	16.5	4.71E+03	2.42E+03	1.19E+03	415.
0.25	76.3	52.4	38.1	26.2	67.1	50.0	33.8	17.1	5.12E+03	2.62E+03	1.29E+03	447.
0.26	79.2	54.4	39.6	27.2	70.0	52.0	35.1	17.6	5.55E+03	2.83E+03	1.39E+03	480.
0.27	82.1	56.4	41.0	28.2	73.0	54.0	36.4	18.2	5.99E+03	3.05E+03	1.49E+03	514.
0.28	85.0	58.4	42.5	29.2	76.0	56.0	37.7	18.8	6.46E+03	3.27E+03	1.60E+03	549.
0.29	87.8	60.4	43.9	30.2	79.1	58.0	38.9	19.4	6.94E+03	3.50E+03	1.71E+03	585.
0.30	90.7	62.4	45.4	31.2	82.1	60.0	40.2	19.9	7.45E+03	3.74E+03	1.82E+03	621.

Table 4.5 (Part 2) Briggs Urban Dispersion Parameters

x, km	Sigma-y, meters				Sigma-z, meters				Sigma-y times Sigma-z			
	A-B	C	D	E-F	A-B	C	D	E-F	A-B	C	D	E-F
0.31	93.6	64.3	46.8	32.2	85.2	62.0	41.5	20.5	7.97E+03	3.99E+03	1.94E+03	659.
0.32	96.4	66.3	48.2	33.1	88.2	64.0	42.8	21.0	8.51E+03	4.24E+03	2.06E+03	697.
0.33	99.3	68.2	49.6	34.1	91.3	66.0	44.1	21.6	9.07E+03	4.50E+03	2.19E+03	737.
0.34	102.	70.2	51.0	35.1	94.5	68.0	45.3	22.1	9.64E+03	4.77E+03	2.31E+03	777.
0.35	105.	72.1	52.4	36.1	97.6	70.0	46.6	22.7	1.02E+04	5.05E+03	2.44E+03	818.
0.36	108.	74.0	53.9	37.0	101.	72.0	47.9	23.2	1.09E+04	5.33E+03	2.58E+03	859.
0.37	111.	76.0	55.3	38.0	104.	74.0	49.1	23.7	1.15E+04	5.62E+03	2.72E+03	902.
0.38	113.	77.9	56.6	38.9	107.	76.0	50.4	24.3	1.21E+04	5.92E+03	2.86E+03	945.
0.39	116.	79.8	58.0	39.9	110.	78.0	51.7	24.8	1.28E+04	6.22E+03	3.00E+03	989.
0.40	119.	81.7	59.4	40.9	114.	80.0	52.9	25.3	1.35E+04	6.54E+03	3.14E+03	1.03E+03
0.41	122.	83.6	60.8	41.8	117.	82.0	54.2	25.8	1.42E+04	6.86E+03	3.29E+03	1.08E+03
0.42	124.	85.5	62.2	42.7	120.	84.0	55.4	26.3	1.49E+04	7.18E+03	3.45E+03	1.13E+03
0.43	127.	87.4	63.6	43.7	123.	86.0	56.7	26.8	1.57E+04	7.51E+03	3.60E+03	1.17E+03
0.44	130.	89.3	64.9	44.6	127.	88.0	57.9	27.3	1.65E+04	7.86E+03	3.76E+03	1.22E+03
0.45	133.	91.1	66.3	45.6	130.	90.0	59.1	27.8	1.72E+04	8.20E+03	3.92E+03	1.27E+03
0.46	135.	93.0	67.6	46.5	133.	92.0	60.4	28.3	1.80E+04	8.56E+03	4.08E+03	1.32E+03
0.47	138.	94.9	69.0	47.4	137.	94.0	61.6	28.8	1.89E+04	8.92E+03	4.25E+03	1.37E+03
0.48	141.	96.7	70.3	48.4	140.	96.0	62.8	29.3	1.97E+04	9.29E+03	4.42E+03	1.42E+03
0.49	143.	98.6	71.7	49.3	144.	98.0	64.1	29.8	2.06E+04	9.66E+03	4.59E+03	1.47E+03
0.50	146.	100.	73.0	50.2	147.	100.	65.3	30.2	2.15E+04	1.00E+04	4.77E+03	1.52E+03
0.55	159.	110.	79.7	54.8	164.	110.	71.3	32.6	2.62E+04	1.21E+04	5.68E+03	1.78E+03
0.60	172.	119.	86.2	59.3	182.	120.	77.3	34.8	3.14E+04	1.42E+04	6.67E+03	2.06E+03
0.65	185.	127.	92.7	63.7	200.	130.	83.2	37.0	3.71E+04	1.66E+04	7.71E+03	2.36E+03
0.70	198.	136.	99.0	68.1	219.	140.	89.1	39.1	4.34E+04	1.91E+04	8.82E+03	2.66E+03
0.75	210.	145.	105.	72.4	238.	150.	94.9	41.2	5.01E+04	2.17E+04	9.98E+03	2.98E+03
0.80	223.	153.	111.	76.6	258.	160.	101.	43.1	5.74E+04	2.45E+04	1.12E+04	3.30E+03
0.85	235.	162.	117.	80.8	277.	170.	106.	45.1	6.52E+04	2.75E+04	1.25E+04	3.64E+03
0.90	247.	170.	123.	84.9	298.	180.	112.	47.0	7.35E+04	3.06E+04	1.38E+04	3.99E+03
0.95	259.	178.	129.	89.0	318.	190.	117.	48.8	8.24E+04	3.38E+04	1.52E+04	4.34E+03
1.00	270.	186.	135.	93.0	339.	200.	123.	50.6	9.18E+04	3.72E+04	1.66E+04	4.70E+03

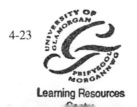

Table 4.5 (Part 3) Briggs Urban Dispersion Parameters

x, km	Sigma-y, meters				Sigma-z, meters				Sigma-y times Sigma-z			
	A-B	C	D	E-F	A-B	C	D	E-F	A-B	C	D	E-F
1.05	282.	194.	141.	96.9	361.	210.	128.	52.3	1.02E+05	4.07E+04	1.81E+04	5.07E+03
1.10	293.	202.	147.	101.	383.	220.	134.	54.1	1.12E+05	4.44E+04	1.96E+04	5.45E+03
1.15	305.	209.	152.	105.	405.	230.	139.	55.7	1.23E+05	4.82E+04	2.11E+04	5.83E+03
1.20	316.	217.	158.	109.	427.	240.	144.	57.4	1.35E+05	5.21E+04	2.27E+04	6.22E+03
1.25	327.	225.	163.	112.	450.	250.	149.	59.0	1.47E+05	5.61E+04	2.44E+04	6.62E+03
1.30	337.	232.	169.	116.	473.	260.	154.	60.6	1.60E+05	6.03E+04	2.60E+04	7.02E+03
1.35	348.	239.	174.	120.	497.	270.	159.	62.1	1.73E+05	6.46E+04	2.78E+04	7.43E+03
1.40	359.	247.	179.	123.	521.	280.	164.	63.6	1.87E+05	6.90E+04	2.95E+04	7.84E+03
1.45	369.	254.	185.	127.	545.	290.	169.	65.1	2.01E+05	7.36E+04	3.13E+04	8.26E+03
1.50	379.	261.	190.	130.	569.	300.	174.	66.6	2.16E+05	7.83E+04	3.31E+04	8.68E+03
1.55	390.	268.	195.	134.	594.	310.	179.	68.0	2.31E+05	8.31E+04	3.49E+04	9.11E+03
1.60	400.	275.	200.	137.	619.	320.	184.	69.4	2.48E+05	8.80E+04	3.68E+04	9.54E+03
1.65	410.	282.	205.	141.	645.	330.	189.	70.8	2.64E+05	9.30E+04	3.87E+04	9.98E+03
1.70	420.	289.	210.	144.	670.	340.	194.	72.2	2.81E+05	9.81E+04	4.06E+04	1.04E+04
1.75	430.	295.	215.	148.	696.	350.	198.	73.5	2.99E+05	1.03E+05	4.26E+04	1.09E+04
1.80	439.	302.	220.	151.	723.	360.	203.	74.9	3.17E+05	1.09E+05	4.46E+04	1.13E+04
1.85	449.	309.	224.	154.	750.	370.	208.	76.2	3.36E+05	1.14E+05	4.66E+04	1.18E+04
1.90	458.	315.	229.	158.	777.	380.	212.	77.5	3.56E+05	1.20E+05	4.86E+04	1.22E+04
1.95	468.	322.	234.	161.	804.	390.	217.	78.7	3.76E+05	1.25E+05	5.07E+04	1.27E+04
2.00	477.	328.	239.	164.	831.	400.	221.	80.0	3.97E+05	1.31E+05	5.28E+04	1.31E+04
2.10	495.	341.	248.	170.	887.	420.	230.	82.5	4.40E+05	1.43E+05	5.70E+04	1.40E+04
2.20	513.	353.	257.	176.	945.	440.	239.	84.9	4.85E+05	1.55E+05	6.14E+04	1.50E+04
2.30	531.	365.	266.	183.	1003.	460.	248.	87.2	5.33E+05	1.68E+05	6.58E+04	1.59E+04
2.40	549.	377.	274.	189.	1062.	480.	256.	89.5	5.83E+05	1.81E+05	7.03E+04	1.69E+04
2.50	566.	389.	283.	194.	1122.	500.	265.	91.8	6.35E+05	1.94E+05	7.48E+04	1.78E+04
2.60	583.	400.	291.	200.	1184.	520.	273.	94.0	6.90E+05	2.08E+05	7.95E+04	1.88E+04
2.70	599.	412.	300.	206.	1246.	540.	281.	96.1	7.47E+05	2.22E+05	8.42E+04	1.98E+04
2.80	615.	423.	308.	212.	1310.	560.	289.	98.2	8.06E+05	2.37E+05	8.89E+04	2.08E+04
2.90	631.	434.	316.	217.	1374.	580.	297.	100.	8.68E+05	2.52E+05	9.37E+04	2.18E+04
3.00	647.	445.	324.	222.	1440.	600.	305.	102.	9.32E+05	2.67E+05	9.86E+04	2.28E+04

Table 4.5 (Part 4) Briggs Urban Dispersion Parameters

x, km	Sigma-y, meters				Sigma-z, meters				Sigma-y times Sigma-z			
	A-B	C	D	E-F	A-B	C	D	E-F	A-B	C	D	E-F
3.10	663.	456.	331.	228.	1506.	620.	312.	104.	9.99E+05	2.83E+05	1.04E+05	2.38E+04
3.20	678.	466.	339.	233.	1574.	640.	320.	106.	1.07E+06	2.98E+05	1.09E+05	2.48E+04
3.30	693.	477.	347.	238.	1642.	660.	328.	108.	1.14E+06	3.15E+05	1.14E+05	2.58E+04
3.40	708.	487.	354.	243.	1712.	680.	335.	110.	1.21E+06	3.31E+05	1.19E+05	2.68E+04
3.50	723.	497.	361.	249.	1782.	700.	342.	112.	1.29E+06	3.48E+05	1.24E+05	2.78E+04
3.60	737.	507.	369.	254.	1853.	720.	349.	114.	1.37E+06	3.65E+05	1.29E+05	2.89E+04
3.70	752.	517.	376.	258.	1925.	740.	357.	116.	1.45E+06	3.82E+05	1.34E+05	2.99E+04
3.80	766.	527.	383.	263.	1998.	760.	364.	117.	1.53E+06	4.00E+05	1.39E+05	3.09E+04
3.90	780.	536.	390.	268.	2072.	780.	371.	119.	1.62E+06	4.18E+05	1.45E+05	3.20E+04
4.00	794.	546.	397.	273.	2147.	800.	378.	121.	1.70E+06	4.37E+05	1.50E+05	3.30E+04
4.10	807.	555.	404.	278.	2222.	820.	384.	123.	1.79E+06	4.55E+05	1.55E+05	3.40E+04
4.20	821.	564.	410.	282.	2299.	840.	391.	124.	1.89E+06	4.74E+05	1.61E+05	3.51E+04
4.30	834.	574.	417.	287.	2376.	860.	398.	126.	1.98E+06	4.93E+05	1.66E+05	3.61E+04
4.40	848.	583.	424.	291.	2454.	880.	404.	128.	2.08E+06	5.13E+05	1.71E+05	3.72E+04
4.50	861.	592.	430.	296.	2533.	900.	411.	129.	2.18E+06	5.32E+05	1.77E+05	3.83E+04
4.60	873.	601.	437.	300.	2613.	920.	417.	131.	2.28E+06	5.52E+05	1.82E+05	3.93E+04
4.70	886.	609.	443.	305.	2693.	940.	424.	133.	2.39E+06	5.73E+05	1.88E+05	4.04E+04
4.80	899.	618.	449.	309.	2774.	960.	430.	134.	2.49E+06	5.93E+05	1.93E+05	4.14E+04
4.90	911.	627.	456.	313.	2856.	980.	436.	136.	2.60E+06	6.14E+05	1.99E+05	4.25E+04
5.00	924.	635.	462.	318.	2939.	1000.	443.	137.	2.72E+06	6.35E+05	2.04E+05	4.36E+04
5.50	984.	676.	492.	338.	3365.	1100.	473.	145.	3.31E+06	7.44E+05	2.33E+05	4.89E+04
6.00	1041.	716.	521.	358.	3810.	1200.	502.	152.	3.97E+06	8.59E+05	2.61E+05	5.43E+04
6.50	1096.	754.	548.	377.	4272.	1300.	530.	159.	4.68E+06	9.80E+05	2.90E+05	5.98E+04
7.00	1149.	790.	575.	395.	4752.	1400.	557.	165.	5.46E+06	1.11E+06	3.20E+05	6.52E+04
7.50	1200.	825.	600.	412.	5000.	1500.	582.	171.	6.00E+06	1.24E+06	3.49E+05	7.07E+04
8.00	1249.	859.	625.	429.	5000.	1600.	607.	178.	6.25E+06	1.37E+06	3.79E+05	7.62E+04
8.50	1297.	891.	648.	446.	5000.	1700.	632.	183.	6.48E+06	1.52E+06	4.09E+05	8.17E+04
9.00	1343.	923.	671.	462.	5000.	1800.	655.	189.	6.71E+06	1.66E+06	4.40E+05	8.73E+04
9.50	1388.	954.	694.	477.	5000.	1900.	678.	195.	6.94E+06	1.81E+06	4.70E+05	9.28E+04
10.00	1431.	984.	716.	492.	5000.	2000.	700.	200.	7.16E+06	1.97E+06	5.01E+05	9.84E+04

Table 4.5   (Part 5)   Briggs Urban Dispersion Parameters

x, km	Sigma-y, meters				Sigma-z, meters				Sigma-y times Sigma-z			
	A-B	C	D	E-F	A-B	C	D	E-F	A-B	C	D	E-F
10.50	1473.	1013.	737.	507.	5000.	2100.	722.	205.	7.37E+06	2.13E+06	5.32E+05	1.04E+05
11.00	1515.	1041.	757.	521.	5000.	2200.	743.	210.	7.57E+06	2.29E+06	5.62E+05	1.10E+05
11.50	1555.	1069.	778.	535.	5000.	2300.	763.	215.	7.78E+06	2.46E+06	5.93E+05	1.15E+05
12.00	1594.	1096.	797.	548.	5000.	2400.	783.	220.	7.97E+06	2.63E+06	6.24E+05	1.21E+05
12.50	1633.	1123.	816.	561.	5000.	2500.	803.	225.	8.16E+06	2.81E+06	6.56E+05	1.26E+05
13.00	1671.	1149.	835.	574.	5000.	2600.	822.	230.	8.35E+06	2.99E+06	6.87E+05	1.32E+05
13.50	1708.	1174.	854.	587.	5000.	2700.	841.	234.	8.54E+06	3.17E+06	7.18E+05	1.38E+05
14.00	1744.	1199.	872.	599.	5000.	2800.	860.	239.	8.72E+06	3.36E+06	7.49E+05	1.43E+05
14.50	1779.	1223.	890.	612.	5000.	2900.	878.	243.	8.90E+06	3.55E+06	7.81E+05	1.49E+05
15.00	1814.	1247.	907.	624.	5000.	3000.	895.	248.	9.07E+06	3.74E+06	8.12E+05	1.54E+05
15.50	1848.	1271.	924.	635.	5000.	3100.	913.	252.	9.24E+06	3.94E+06	8.44E+05	1.60E+05
16.00	1882.	1294.	941.	647.	5000.	3200.	930.	256.	9.41E+06	4.14E+06	8.75E+05	1.66E+05
16.50	1915.	1317.	958.	658.	5000.	3300.	947.	260.	9.58E+06	4.35E+06	9.07E+05	1.71E+05
17.00	1948.	1339.	974.	670.	5000.	3400.	964.	264.	9.74E+06	4.55E+06	9.38E+05	1.77E+05
17.50	1980.	1361.	990.	681.	5000.	3500.	980.	268.	9.90E+06	4.76E+06	9.70E+05	1.83E+05
18.00	2011.	1383.	1006.	691.	5000.	3600.	996.	272.	1.01E+07	4.98E+06	1.00E+06	1.88E+05
18.50	2043.	1404.	1021.	702.	5000.	3700.	1012.	276.	1.02E+07	5.20E+06	1.03E+06	1.94E+05
19.00	2073.	1425.	1037.	713.	5000.	3800.	1028.	280.	1.04E+07	5.42E+06	1.07E+06	1.99E+05
19.50	2104.	1446.	1052.	723.	5000.	3900.	1043.	284.	1.05E+07	5.64E+06	1.10E+06	2.05E+05
20.00	2133.	1467.	1067.	733.	5000.	4000.	1058.	287.	1.07E+07	5.87E+06	1.13E+06	2.11E+05
21.00	2192.	1507.	1096.	753.	5000.	4200.	1088.	295.	1.10E+07	6.33E+06	1.19E+06	2.22E+05
22.00	2249.	1546.	1124.	773.	5000.	4400.	1117.	302.	1.12E+07	6.80E+06	1.26E+06	2.33E+05
23.00	2305.	1584.	1152.	792.	5000.	4600.	1146.	309.	1.15E+07	7.29E+06	1.32E+06	2.45E+05
24.00	2359.	1622.	1179.	811.	5000.	4800.	1173.	316.	1.18E+07	7.78E+06	1.38E+06	2.56E+05
25.00	2412.	1658.	1206.	829.	5000.	5000.	1200.	322.	1.21E+07	8.29E+06	1.45E+06	2.67E+05
26.00	2464.	1694.	1232.	847.	5000.	5000.	1227.	329.	1.23E+07	8.47E+06	1.51E+06	2.79E+05
27.00	2515.	1729.	1258.	865.	5000.	5000.	1253.	335.	1.26E+07	8.65E+06	1.58E+06	2.90E+05
28.00	2565.	1764.	1283.	882.	5000.	5000.	1279.	342.	1.28E+07	8.82E+06	1.64E+06	3.01E+05
29.00	2614.	1797.	1307.	899.	5000.	5000.	1304.	348.	1.31E+07	8.99E+06	1.70E+06	3.13E+05
30.00	2663.	1831.	1331.	915.	5000.	5000.	1328.	354.	1.33E+07	9.15E+06	1.77E+06	3.24E+05

Table 4.5 (Part 6) Briggs Urban Dispersion Parameters

x, km	Sigma-y, meters				Sigma-z, meters				Sigma-y times Sigma-z			
	A-B	C	D	E-F	A-B	C	D	E-F	A-B	C	D	E-F
31.00	2710.	1863.	1355.	932.	5000.	5000.	1352.	360.	1.35E+07	9.32E+06	1.83E+06	3.35E+05
32.00	2757.	1895.	1378.	948.	5000.	5000.	1376.	366.	1.38E+07	9.48E+06	1.90E+06	3.47E+05
33.00	2802.	1927.	1401.	963.	5000.	5000.	1399.	371.	1.40E+07	9.63E+06	1.96E+06	3.58E+05
34.00	2847.	1958.	1424.	979.	5000.	5000.	1422.	377.	1.42E+07	9.79E+06	2.02E+06	3.69E+05
35.00	2892.	1988.	1446.	994.	5000.	5000.	1445.	383.	1.45E+07	9.94E+06	2.09E+06	3.81E+05
36.00	2936.	2018.	1468.	1009.	5000.	5000.	1467.	388.	1.47E+07	1.01E+07	2.15E+06	3.92E+05
37.00	2979.	2048.	1489.	1024.	5000.	5000.	1489.	394.	1.49E+07	1.02E+07	2.22E+06	4.03E+05
38.00	3021.	2077.	1511.	1039.	5000.	5000.	1511.	399.	1.51E+07	1.04E+07	2.28E+06	4.15E+05
39.00	3063.	2106.	1532.	1053.	5000.	5000.	1532.	404.	1.53E+07	1.05E+07	2.35E+06	4.26E+05
40.00	3104.	2134.	1552.	1067.	5000.	5000.	1553.	410.	1.55E+07	1.07E+07	2.41E+06	4.37E+05
41.00	3145.	2162.	1573.	1081.	5000.	5000.	1574.	415.	1.57E+07	1.08E+07	2.48E+06	4.49E+05
42.00	3186.	2190.	1593.	1095.	5000.	5000.	1594.	420.	1.59E+07	1.10E+07	2.54E+06	4.60E+05
43.00	3225.	2217.	1613.	1109.	5000.	5000.	1615.	425.	1.61E+07	1.11E+07	2.60E+06	4.71E+05
44.00	3265.	2244.	1632.	1122.	5000.	5000.	1635.	430.	1.63E+07	1.12E+07	2.67E+06	4.83E+05
45.00	3304.	2271.	1652.	1136.	5000.	5000.	1654.	435.	1.65E+07	1.14E+07	2.73E+06	4.94E+05
46.00	3342.	2298.	1671.	1149.	5000.	5000.	1674.	440.	1.67E+07	1.15E+07	2.80E+06	5.05E+05
47.00	3380.	2324.	1690.	1162.	5000.	5000.	1693.	445.	1.69E+07	1.16E+07	2.86E+06	5.17E+05
48.00	3418.	2350.	1709.	1175.	5000.	5000.	1712.	449.	1.71E+07	1.17E+07	2.93E+06	5.28E+05
49.00	3455.	2375.	1727.	1188.	5000.	5000.	1731.	454.	1.73E+07	1.19E+07	2.99E+06	5.39E+05
50.00	3491.	2400.	1746.	1200.	5000.	5000.	1750.	459.	1.75E+07	1.20E+07	3.06E+06	5.51E+05
51.00	3528.	2425.	1764.	1213.	5000.	5000.	1768.	463.	1.76E+07	1.21E+07	3.12E+06	5.62E+05
52.00	3564.	2450.	1782.	1225.	5000.	5000.	1787.	468.	1.78E+07	1.23E+07	3.18E+06	5.73E+05
53.00	3600.	2475.	1800.	1237.	5000.	5000.	1805.	473.	1.80E+07	1.24E+07	3.25E+06	5.85E+05
54.00	3635.	2499.	1817.	1249.	5000.	5000.	1823.	477.	1.82E+07	1.25E+07	3.31E+06	5.96E+05
55.00	3670.	2523.	1835.	1262.	5000.	5000.	1841.	482.	1.83E+07	1.26E+07	3.38E+06	6.07E+05
56.00	3705.	2547.	1852.	1273.	5000.	5000.	1858.	486.	1.85E+07	1.27E+07	3.44E+06	6.19E+05
57.00	3739.	2570.	1869.	1285.	5000.	5000.	1876.	490.	1.87E+07	1.29E+07	3.51E+06	6.30E+05
58.00	3773.	2594.	1886.	1297.	5000.	5000.	1893.	495.	1.89E+07	1.30E+07	3.57E+06	6.41E+05
59.00	3807.	2617.	1903.	1309.	5000.	5000.	1910.	499.	1.90E+07	1.31E+07	3.64E+06	6.53E+05
60.00	3840.	2640.	1920.	1320.	5000.	5000.	1927.	503.	1.92E+07	1.32E+07	3.70E+06	6.64E+05

Table 4.5 (Part 7) Briggs Urban Dispersion Parameters

x, km	Sigma-y, meters				Sigma-z, meters				Sigma-y times Sigma-z			
	A-B	C	D	E-F	A-B	C	D	E-F	A-B	C	D	E-F
61.00	3873.	2663.	1937.	1331.	5000.	5000.	1944.	507.	1.94E+07	1.33E+07	3.76E+06	6.76E+05
62.00	3906.	2685.	1953.	1343.	5000.	5000.	1961.	512.	1.95E+07	1.34E+07	3.83E+06	6.87E+05
63.00	3939.	2708.	1969.	1354.	5000.	5000.	1977.	516.	1.97E+07	1.35E+07	3.89E+06	6.98E+05
64.00	3971.	2730.	1985.	1365.	5000.	5000.	1994.	520.	1.99E+07	1.36E+07	3.96E+06	7.10E+05
65.00	4003.	2752.	2001.	1376.	5000.	5000.	2010.	524.	2.00E+07	1.38E+07	4.02E+06	7.21E+05
66.00	4035.	2774.	2017.	1387.	5000.	5000.	2026.	528.	2.02E+07	1.39E+07	4.09E+06	7.32E+05
67.00	4066.	2796.	2033.	1398.	5000.	5000.	2042.	532.	2.03E+07	1.40E+07	4.15E+06	7.44E+05
68.00	4098.	2817.	2049.	1409.	5000.	5000.	2058.	536.	2.05E+07	1.41E+07	4.22E+06	7.55E+05
69.00	4129.	2838.	2064.	1419.	5000.	5000.	2074.	540.	2.06E+07	1.42E+07	4.28E+06	7.66E+05
70.00	4160.	2860.	2080.	1430.	5000.	5000.	2089.	544.	2.08E+07	1.43E+07	4.35E+06	7.78E+05
71.00	4190.	2881.	2095.	1440.	5000.	5000.	2105.	548.	2.10E+07	1.44E+07	4.41E+06	7.89E+05
72.00	4221.	2902.	2110.	1451.	5000.	5000.	2120.	552.	2.11E+07	1.45E+07	4.47E+06	8.00E+05
73.00	4251.	2922.	2125.	1461.	5000.	5000.	2136.	556.	2.13E+07	1.46E+07	4.54E+06	8.12E+05
74.00	4281.	2943.	2140.	1472.	5000.	5000.	2151.	559.	2.14E+07	1.47E+07	4.60E+06	8.23E+05
75.00	4311.	2963.	2155.	1482.	5000.	5000.	2166.	563.	2.16E+07	1.48E+07	4.67E+06	8.35E+05
76.00	4340.	2984.	2170.	1492.	5000.	5000.	2181.	567.	2.17E+07	1.49E+07	4.73E+06	8.46E+05
77.00	4369.	3004.	2185.	1502.	5000.	5000.	2196.	571.	2.18E+07	1.50E+07	4.80E+06	8.57E+05
78.00	4399.	3024.	2199.	1512.	5000.	5000.	2211.	574.	2.20E+07	1.51E+07	4.86E+06	8.69E+05
79.00	4428.	3044.	2214.	1522.	5000.	5000.	2225.	578.	2.21E+07	1.52E+07	4.93E+06	8.80E+05
80.00	4456.	3064.	2228.	1532.	5000.	5000.	2240.	582.	2.23E+07	1.53E+07	4.99E+06	8.91E+05
81.00	4485.	3083.	2242.	1542.	5000.	5000.	2255.	585.	2.24E+07	1.54E+07	5.06E+06	9.03E+05
82.00	4513.	3103.	2257.	1551.	5000.	5000.	2269.	589.	2.26E+07	1.55E+07	5.12E+06	9.14E+05
83.00	4542.	3122.	2271.	1561.	5000.	5000.	2283.	593.	2.27E+07	1.56E+07	5.18E+06	9.25E+05
84.00	4570.	3142.	2285.	1571.	5000.	5000.	2298.	596.	2.28E+07	1.57E+07	5.25E+06	9.37E+05
85.00	4598.	3161.	2299.	1580.	5000.	5000.	2312.	600.	2.30E+07	1.58E+07	5.31E+06	9.48E+05
86.00	4625.	3180.	2313.	1590.	5000.	5000.	2326.	603.	2.31E+07	1.59E+07	5.38E+06	9.59E+05
87.00	4653.	3199.	2326.	1599.	5000.	5000.	2340.	607.	2.33E+07	1.60E+07	5.44E+06	9.71E+05
88.00	4680.	3218.	2340.	1609.	5000.	5000.	2354.	610.	2.34E+07	1.61E+07	5.51E+06	9.82E+05
89.00	4708.	3236.	2354.	1618.	5000.	5000.	2367.	614.	2.35E+07	1.62E+07	5.57E+06	9.93E+05
90.00	4735.	3255.	2367.	1628.	5000.	5000.	2381.	617.	2.37E+07	1.63E+07	5.64E+06	1.00E+06

Table 4.5 (Part 8) Briggs Urban Dispersion Parameters

x, km	Sigma-y, meters				Sigma-z, meters				Sigma-y times Sigma-z			
	A-B	C	D	E-F	A-B	C	D	E-F	A-B	C	D	E-F
91.00	4762.	3274.	2381.	1637.	5000.	5000.	2395.	621.	2.38E+07	1.64E+07	5.70E+06	1.02E+06
92.00	4788.	3292.	2394.	1646.	5000.	5000.	2408.	624.	2.39E+07	1.65E+07	5.77E+06	1.03E+06
93.00	4815.	3310.	2408.	1655.	5000.	5000.	2422.	628.	2.41E+07	1.66E+07	5.83E+06	1.04E+06
94.00	4842.	3329.	2421.	1664.	5000.	5000.	2435.	631.	2.42E+07	1.66E+07	5.90E+06	1.05E+06
95.00	4868.	3347.	2434.	1673.	5000.	5000.	2449.	634.	2.43E+07	1.67E+07	5.96E+06	1.06E+06
96.00	4894.	3365.	2447.	1682.	5000.	5000.	2462.	638.	2.45E+07	1.68E+07	6.02E+06	1.07E+06
97.00	4920.	3383.	2460.	1691.	5000.	5000.	2475.	641.	2.46E+07	1.69E+07	6.09E+06	1.08E+06
98.00	4946.	3400.	2473.	1700.	5000.	5000.	2488.	644.	2.47E+07	1.70E+07	6.15E+06	1.10E+06
99.00	4972.	3418.	2486.	1709.	5000.	5000.	2501.	648.	2.49E+07	1.71E+07	6.22E+06	1.11E+06
100.00	4998.	3436.	2499.	1718.	5000.	5000.	2514.	651.	2.50E+07	1.72E+07	6.28E+06	1.12E+06

CHAPTER 5.

PUTTING GAUSSIAN METHODS INTO PERSPECTIVE

Gaussian methods can be expected to apply to any situation where the distribution of the short term velocities in that dimension (horizontal or vertical) can be expected to be well represented by a Gaussian or normal distribution over the selected averaging time (usually an hour). In general, this is the case for the horizontal dimension, although there are periods, especially under stable conditions, when there is a gradual change in the wind direction over the one-hour period that the horizontal dispersion would not be well represented by the Gaussian distribution.

5.1 Where Gaussian Methods Are Less Applicable

There are two situations where the Gaussian distribution does not represent well the distribution of vertical velocities. These are for convective situations and for groundlevel releases.

5.2 Convective Situations

The area occupied by thermals under the extreme convective situation is on the order of 35 to 40 percent. In the center of the thermals, at heights midway between the ground and the mixing height, the upward vertical velocities are about 2 m s^{-1}. To complete the mass continuity the downward velocities over the remaining 60 to 65 percent of the area are on the order of 1 m s^{-1}. This results in a non-Gaussian vertical velocity distribution with few but higher-velocity upward motions and more but lower-velocity downward motions (Wyngaard, 1988; Weil, 1988). This can be simulated properly by assuming the approximate distribution under these conditions and estimating the vertical dispersion using this distribution. This approach is commonly called the PDF or probability density function (Weil and Brower, 1984; Berkowicz, Olesen, and Torp, 1986; Weil and Corio, 1988).

A model for elevated releases that incorporates the PDF is HPDM (Hanna and Paine, 1989; Hanna and Chang, 1990) with support from EPRI (Electric Power Research Institute).

The major reason that the P-G curve for the A Pasquill stability has such an upsweep at the greater distances is that the σ_z is calculated from the groundlevel concentrations measured from field data using an assumption that there is a Gaussian vertical distribution.

5.3 Groundlevel Releases

Near the ground the increase of wind speed with height due to the surface friction is such that it is not possible to select a single wind speed which will be appropriate as a dilution speed. The turbulence is not homogeneous in the vertical due to the presence of the surface. Therefore Gaussian techniques are not appropriate (Gryning, van Ulden, and Larsen, 1983). An alternative technique using a K-theory approach is more appropriate (Nieuwstadt and van Ulden, 1978; Gryning et al.,1987). This technique uses a constant K that represents a vertical diffusivity.

The plume centerline generally does not stay at the same height under convective conditions. For elevated positions the plume descends. For groundlevel releases the plume lifts off the ground. This was verified by laboratory studies (Deardorff and Willis, 1975; Willis and Deardorff, 1978) and numerical simulations (Lamb, 1982). The descent of an elevated release and the rising of a surface release were verified in the field by the CONDORS experiment (Moninger et al., 1983; Eberhard et al., 1985; Briggs et al., 1986) conducted near Boulder in 1982 and 1983.

CHAPTER 6.

USING COMPUTERS FOR DISPERSION ESTIMATES

What is a dispersion model? It was stated in Chapter 1 that it is when a computer is used for the repetitious solution of these equations that we refer to the calculations as a dispersion model.

The primary inputs to a dispersion model consist of emission information and meteorological data. See Figure 1.4. The emission information consists of the coordinates for the location of the source, the physical stack height, the inside stack top diameter, the stack exit velocity, and the stack temperature. The meteorological parameters for hourly periods are Pasquill stability class, wind direction, wind speed, temperature, and mixing height.

The calculations by the model are organized so that there is an inner loop that is executed for each receptor (Figure 6.1). Two major calculations are accomplished within this loop. First, by using analytical geometry, the downwind distance and the crosswind distance of the receptor from the source are determined. The other calculation is solving a dispersion equation to determine the pollutant concentration at this receptor from a single source.

Outside of this inner loop is a second loop that is executed for each source. The major computation that is done within this loop is the calculation of plume rise and effective plume height for this source for the meteorological condition for this hour.

Outside the above loop is a third loop that is executed for each hour. Prior to entry to this loop, general setup of the run is accomplished, including selection of various options such as using urban or rural dispersion parameters and the reading of source information and provision for receptors. Access to the appropriate meteorology for this hour is accomplished within this loop.

Although most short-term models can be executed with all of the data in a single run-stream reading from one input file, most of these models will also allow the meteorological data to be read from a separate file. This latter mode is usually applied for normal short-term runs that are making hour-by-hour simulations for an entire year.

Although the basic output from most models consist of hourly concentrations at the receptors, many of the models contain considerable code to do "bookkeeping." Averages over period longer than an hour are calculated and the highest ones retained throughout the model run. When the run is completed, it is possible to print tables that provide highest and second highest concentrations for each receptor for averaging times that include one-hour, three-hour, eight-hour, and 24-hours. Also, the average concentrations over the period of simulation, frequently a year, are in tabular form.

Information about the diskette furnished with this volume is discussed in the next chapter.

DO FOR EACH SIMULATED METEOROLOGICAL
PERIOD (HOUR)

DO FOR EACH SOURCE

Calculate Effective Plume Height

for this Source

(# of hours times # of sources)

DO FOR EACH RECEPTOR

Analytic Geometry

to Determine Downwind and

Crosswind Distances

of Receptor from Source.

Determine Concentration by

Solving Dispersion Equation

(# of hours times # of sources

times # of receptors)

Figure 6.1 The calculations made by a short-term model.

CHAPTER 7.

USE OF THE COMPUTER DISKETTE

The computer diskette included with this workbook provides executable programs that will allow you to make a variety of dispersion calculations. The programs are designed to give the user some familiarity with various techniques used to make dispersion estimates. An understanding of the sensitivity of the concentration estimates to various changes in the values of the input parameters can be gained by changing the value of just one parameter and running the program again. It is not intended to make a series of computations to provide a complete analysis of a situation, as some regulatory screening and refined models can provide.

Two executable programs are provided. WKBK2.EXE is the primary program. SUPL.EXE is a supplementary program that will calculate parameter values that are required for some options selected in WKBK2. To properly execute the program, it is necessary to turn the "caps lock" on. A READ.ME file is on the diskette. It is suggested that the reader print the file or use his editor to read the file before trying to use the programs.

7.1 Installation of the Program

The full screen menus in the program were programmed using FORTRAN. In order for the full screen menu to function correctly, the extended screen and keyboard control device driver, ANSI.SYS must be installed by placing the following statement in the configuration file, CONFIG.SYS:

DEVICE=ANSI.SYS

Also, the file ANSI.SYS must exist in the root directory or in another directory that is named in the search path of the AUTOEXEC.BAT file. If you do not already have an ANSI.SYS in such a directory, read this file from the floppy into an appropriate directory.

It is easy to get ready to run the programs from a separate directory on your hard drive by using the program INSTALL.BAT furnished on the diskette. While in the drive that you are using to read the diskette, type INSTALL furnishing the values of three parameters on the command line separated by spaces: 1) The drive you want to use (normally a hard drive), 2) the directory you want established for these programs, and 3) the drive you are using to read the diskette. An example of this command is:

INSTALL C WKBK2 A

Following the execution of install, the typing of START will allow easy execution of either of the programs.

7.2 Running the Program - WKBK2

The program can be run either by using the START.BAT program furnished on the diskette or by typing WKBK2 and [return]. The user can create a report, in addition to complete information displayed on the screen as the run is made or can just obtain output displayed on the screen as the run progresses. Input to the program is accomplished by entries provided on menus or by answering questions. There are six menus that are built into the program. These are:

1) Report Header Menu
2) Main Menu
3) Source and Options Menu
4) Meteorology Menu
5) Receptor and Isopleths Menu
6) Dispersion Menu.

The Report Header Menu is only called if the user decides to generate a report. If this is the case, three lines of title are entered. The number of lines that the user's printer prints before advancing to the next page and the number of lines desired on each page of the report are required. Some experimentation may be required with entry of these values to get the desired result. Also, the page number to appear on the first page of the report is entered.

The opportunity is available to cycle through each entry of each menu as many times as is required to put in those values that are desired by the user. Exit from each menu is by answering the question, "Do you want to change any of the values on this menu?" with an "N". This is the first question on each menu, and after cycling through each possible entry, the menu returns to this question.

The Main Menu allows for acceptance of six possible digits, 1-6. An entry of 5 allows sequencing through all menus. Entries of 1 through 4 allows entry beginning with different menus. Entries of 1 to 4 are normally used only to change one or a few parameter values after sequencing through all menus and obtaining some resulting calculations. The Main Menu is also the point where exit from the program occurs. An entry of 6 will cause a programmed exit.

The Source and Options Menu provides for entry of emission rate and the physical parameters for the source: height, diameter, exit velocity, and exit temperature. This program allows the consideration of only one source. This menu also provides for use of two optional features: buoyancy-induced dispersion or building downwash. Selection of entering effective height or of using two different plume rise calculations is also by this menu. If the '91 plume rise is selected, entries are made for additional parameters of surface heat flux, station pressure, and friction velocity. See the description of the Supplementary program, SUPL, below for information on how to obtain these values.

The Meteorology Menu provides for entry of mixing height and the number of layers for which meteorological data will be entered. For the cases with stable conditions at groundlevel the mixing height is undefined, since the mixing height is the depth of the vigorously mixed layer. For these cases a large value can be entered in this menu for the mixing height. Also, selection is made for entry of temperature at each height or change of potential temperature with height, $d\theta/dz$, for each layer. Heights are specified by the user for each layer. For stack top and each of these layer heights, entry is made of wind speeds and either temperature or potential temperature with height.

The Receptor and Isopleths Menu allows for selection of five different receptor possibilities. One of these is to determine maximum groundlevel concentrations. A second method is to determine the concentrations directly downwind of the source at distances entered by the user. For this method (and methods three through five) it is possible to have the receptors at groundlevel or to have the receptors at a height above the ground. These are referred to as flagpole receptors. A third method, "Coordinates Relative to the Source", allows entry of downwind and crosswind distances of each receptor from the source. The fourth method, "Polar Coordinates Relative to Source", allows entry of a distance and an azimuth of the receptor from the source. The fifth method allows the entry of coordinates of the source and of each receptor. Since the distances in the model are in kilometers, a "scale" factor can be entered. This is chosen so that multiplication of the user's coordinates by this scale converts them to kilometers.

The Dispersion Menu allows for calculation with five dispersion techniques. Three of these, Briggs Urban, Briggs Rural, and Pasquill-Gifford require the Pasquill stability class, which is entered by the user. The Pasquill-Gifford dispersion parameters, usually assumed appropriate for rural conditions, were previously discussed in Chapter 2. The user can obtain assistance in approximating the Pasquill Stability Class by using the Supplementary program described below. The Briggs Urban dispersion parameters, representative of urban conditions, were discussed in Chapter 4. The Brookhaven method uses the specification of the Brookhaven class (Singer and Smith, 1953). The other method, "Fluctuations", requires the specification of both the horizontal and vertical fluctuations. The method used here to convert the hourly horizontal fluctuations, the standard deviation of azimuth angle of the wind, to σ_y and to convert the hourly vertical fluctuations, the standard deviation of elevation angle of the wind, to σ_z are those of Irwin (1983). These are assumed to be for the effective plume height. In addition, the user must specify whether the atmosphere is stable or unstable.

At the conclusion of the Dispersion Menu, the model begins making the computations. If additional entry of parameter values are required, questions will be displayed on the screen asking for information. At the conclusion of the computations, the model will always return to the Main Menu. Selections can be made to make additional computations. In order to exit from the program it is necessary to enter a "6" for the action to be taken, followed by an "N" in reply to, "Do you want to change the selected action?". This will cause a programmed end to the run.

7.3 The Supplementary Program - SUPL

The supplementary program is provided to assist in determining the variables that are required for the plume rise methodology referred to as "91" in the WKBK2 program.

These are surface sensible heat flux and surface friction velocity. The program can also be used to estimate the Pasquill stability class. To use the supplementary program type SUPL and [return] or use the START program.

In order to calculate the surface sensible heat flux a method of Smith (1973) is used which requires total cloud amount and the elevation angle of the sun (during the daytime). To determine the elevation angle, the latitude and longitude of the site are required along with the year, month, day, and hour. Methodology to calculate the time of sunrise, sunset, and the elevation angle of the sun is abstracted and used from the meteorological processor RAMMET (see pp 5 to 6 and 35 to 37 of Vol. I, and pp 36 to 37 of Vol. II of Turner and Novak, 1978). The calculated time zone, time of sunrise and sunset, the time of meridian passage, and the elevation angle of the sun are informational variable values that are displayed.

When the elevation angle is positive, the calculated elevation angle along with the input cloud amount in tenths are then used to estimate the surface sensible heat flux. When the sun is below the horizon, the cloud amount and the 10-meter wind speed is used to estimate the surface sensible heat flux using procedures developed by Smith (1983).

To obtain the friction velocity, during positive sensible heat flux, the 10-meter wind speed, the surface roughness length, and the surface sensible heat flux are used. Surface similarity considerations are used in a methodology by Farmer (1983) which makes use of methods of Dyer (1974) and Huang (1979) related to the theoretical wind profile for unstable conditions. In addition to obtaining u*, the Monin-Obukhov length is an added output.

When the surface sensible heat flux is negative, the cloud cover, 10-meter wind speed, surface roughness, surface pressure, and temperature are used to estimate the friction velocity and Monin-Obukhov length using procedures of Hanna and Chang (1992) used in their SIGPRO (the meteorological processor for the HPDM). This uses the method for u* from Weil and Brower (1983) and the use of an intermediate variable $\theta*$ using Holtslag and van Ulden (1983).

The resulting estimated values of heat flux and surface friction velocity can be used as inputs to WKBK2 to calculate plume rise with the '91 method.

Since, in obtaining the above information for plume rise, the cloud cover and 10-meter wind speed were required, and the surface sensible heat flux was determined, it was thought useful to employ the method of F. B. Smith (1973) to determine his stability parameter, P. The Pasquill stability class can be estimated from P, so both of these parameters are also provided as output.

All input that is required for these estimates is entered on a single input menu which works similarly to those in WKBK2. One can go through the entries as many times as needed and then answer "N" to the question, "Do you want to change any of the values on this menu?". Since the capability exists to produce a report, it is possible to process information for several hours of the same day without needing to repeatedly enter data for many of the variables.

CHAPTER 8.

EXAMPLE PROBLEMS

The following example problems and their solutions illustrate the application of most of the techniques and equations presented in the previous chapters of this workbook. Emission rates and stated concentrations of interest may or may not be of relevance to actual emission rates and air quality concentrations of real interest and concern and are used only for illustration.

PROBLEM 1.

Three grams of an air pollutant are emitted from a 12 meter stack that has a diameter of 0.5 meter. The exit velocity is 15 m s^{-1} and the exit temperature is 315 °K. The local environment is considered rural.

1a. On an afternoon when the Pasquill stability class is judged to be B, what is the plume rise if the wind speed at stack top is 4 m s^{-1}? Assume that the ambient air temperature is 20 °C (293 °K).

1b. For these atmospheric conditions and the plume rise found in part a, use Figure 2.5 to approximate the distance to the maximum groundlevel concentration and the maximum concentration.

1c. Using the distance to the maximum approximated in part b, find the Pasquill-Gifford dispersion parameters from Table 2.5. Use equation 2.3 to determine the groundlevel concentration beneath the plume centerline.

1d. Using the WKBK2 program on the floppy diskette, and the conditions specified above, determine the distance to the maximum concentration and the maximum concentration. If the distance to the maximum that you approximated from Figure 2.5 differs from the distance found by running the program, use the program to also determine the groundlevel concentration at this distance that you found in part b.

SOLUTION:

1a. Since the ratio of exit velocity to wind speed is 15 to 4, equal to 3.75 and thus greater than 1.5, there is no stack tip downwash. Using equation 3.3, the buoyancy flux, F, is 0.642. Using equation 3.4 the plume rise is calculated to be 3.84 meters. This gives an effective plume height of 15.84 meters.

1b. Using Figure 2.5 and interpolating between the effective plume heights of 15 and 20 meters along the B stability line, the distance of maximum concentration is approximated to be at a downwind distance of 0.11 km. and the $\chi u/Q_{max}$ is about 5 x 10^{-4} m^{-2}.

Multiplying this by a Q/u of $3/4 = 0.75$ yields a maximum groundlevel concentration, χ_{max}, of 3.75×10^{-4} g m^{-3} or 375 µg m^{-3}.

1c. At this distance of 0.11 km, Table 2.5 gives $\sigma_y = 21.0$ and $\sigma_z = 11.6$ for Pasquill class B. Using equation 2.3 gives $\chi_{max} = 3.86 \times 10^{-4}$ g m^{-3} or 386 µg m^{-3}.

1d. Running the program WKBK2 for these conditions gives a distance to the maximum of 0.107 km and a maximum concentration the same as calculated above, 3.86×10^{-4} g m^{-3}. Using the program to calculate the concentration at the downwind distance of 0.11 km gives a maximum concentration of 3.85×10^{-4} g m^{-3}.

NOTE: Frequently in a real situation a stack of only 12 meters high will have buildings in the vicinity that are likely to cause building downwash. Discussion of building downwash is not included in this workbook, although it will be noted that there is a building downwash option included in the program WKBK2 that provides some calculations for this effect.

PROBLEM 2.

For the same source and conditions of Problem 1, except that the stability class is judged to be Pasquill class D stability, what is the plume rise, distance to the point of maximum groundlevel concentration, and maximum groundlevel concentration?

SOLUTION:

Since the same source is involved, the buoyancy flux, F, will remain unchanged. Since the wind speed is also unchanged and equation 3.4 is for unstable-neutral conditions and D stability is considered neutral, the plume rise will also remain unchanged and will be 3.84 meters. The effective plume height will be at 15.84 meters. Using Figure 2.5 the distance to the point of maximum groundlevel concentration is about 0.28 km. and the $\chi u/Q_{max}$ is about 5×10^{-4} m^{-2}. This is the same as in problem 1. Multiplying by emission rate and dividing by wind speed will give a χ_{max}, of 3.75×10^{-4} g m^{-3} or 375 µg m^{-3}.

At the downwind distance of 0.28 km, for D stability, Table 2.5 gives $\sigma_y = 22.6$ m and $\sigma_z = 12.1$. Using equation 2.3 gives $\chi_{max} = 3.70 \times 10^{-4}$ g m^{-3} or 370 µg m^{-3}. Using the program WKBK2, χ_{max} is calculated to be 3.76×10^{-4} g m^{-3} at a distance of 0.27 km. The program calculates the concentration at 0.28 km to be 3.75×10^{-4} g m^{-3} or 375 µg m^{-3}.

PROBLEM 3.

A steam plant in a rural area with an approximate surface roughness, z_o, of 0.3 m is located at 32.9 north latitude and 96.9 west longitude. The plant emits 300 g s^{-1} from a 105-meter stack that has a stack top inside diameter of 5.5 meters with stack gas exit velocity of 20 m s^{-1} and stack gas exit temperature of 400 °K. If on November 7, 1992 for the hour ending at 1500, the wind and vertical potential temperature change are as given in the following table, the cloud cover is 9/10, and the mixing height is at 500 meters, what is the effective plume height, the distance to maximum concentration, and the maximum concentration? Assume that the air temperature is 293 °K.

Height m	Wind Speed $m\,s^{-1}$	$\partial\theta/\partial z$ $°K\,m^{-1}$
10.	4.0	-0.0005
105.	5.7	-0.0005
300.	6.7	-0.0005
500.	7.	0.0150
1200.	7.	0.0150
1500.	7.	

SOLUTION:

Using the SUPL program on the diskette, it is found that for this latitude, time of year, time of day, and cloud cover, that the Pasquill stability class is estimated to be D. From equation 3.3 the buoyancy flux is 397. Therefore, for unstable-neutral conditions equation 3.5 is used to estimate plume rise. The plume rise is 246.2 meters and the effective plume height is 351.2 meters. From Figure 2.5 it is estimated that the maximum concentration will occur at a distance of about 30 km from the source. Using the WKBK2 program on the diskette, the maximum concentration is 19.3 µg m^{-3} and the distance to maximum concentration is 26.964 km.

Using the surface heat flux of 25.5 watts and the friction velocity, u*, of 0.468 m s^{-1} obtained from running SUPL, the '91 method (plume rise method 3) of calculating plume rise through layers, and the same data, gives a plume top 507.4 meters above the stack top, 112.4 meters above the mixing height. Assuming a uniform distribution of the plume from bottom to top, a top hat distribution, it is calculated that 0.67 of the plume is below the mixing height. Using the mean height of that part of the plume below the mixing height, 387.1 m, the maximum concentration of 11.1 µg m^{-3} is found at 32.674 km.

PROBLEM 4.

For the same time and source as in Problem 3, if the surface (10 m) wind speed is 8 m s^{-1} and the wind with height is as given in the following table, what is the plume height, the maximum concentration, and the distance to the maximum concentration?

Height m	Wind Speed m s^{-1}	$\partial\theta/\partial z$ °K m^{-1}
10.	8.0	-0.0005
105.	11.4	-0.0005
300.	13.3	-0.0005
500.	14.5	0.0150
1200.	13.0	0.0150
1500.	13.0	

SOLUTION:

Using the SUPL program, the stability and heat flux are the same as for Problem 3, but the friction velocity, u^*, is calculated to be 0.916 m s^{-1}. Equation 3.5 gives 123 m for the plume rise, just half that of the situation in Problem 3. This gives a plume height of 228 m. Figure 2.5 shows a distance to maximum concentration of about 10 km. Using WKBK2 the maximum concentration is calculated to be 27.6 µg m^{-3} at a distance 10.977 km.

Calculating plume rise through layers, the rise for this wind distribution is found to be 93 m and the plume height to be 198 m. Figure 2.5 indicates that the maximum would occur in the vicinity of 9 km. The maximum $\chi u/Q$ is estimated to be about 1.5 x 10^{-6}. Multiplying this by Q/u or 300./11.4 since the wind at stack top is stated to be 11.4 m s^{-1} yields an estimated maximum concentration of 39.5 µg m^{-3}. Using the program WKBK2, the maximum concentration is found to be 39.3 µg m^{-3} at a distance of 8.920 km.

PROBLEM 5.

An industrial source is in a rural location having a surface roughness of 0.4 m at 53° N latitude and 3° W longitude. It has a 30 m stack with a top inside diameter of 2.6 m. Under its normal operating capacity it emits 50 g s^{-1} of fine particulate matter with mass median diameter of less than 10 µm (PM10) at an exit velocity of 20 m s^{-1} and with an exit temperature of 325°K. On June 18th at 1100 with clear skies, the wind speed at 10 meters above ground is 3 m s^{-1}, and the wind at 30 meters above ground is 3.5 m s^{-1}. The surface pressure is 997 mb, the ambient air temperature is 20 °C, and the mixing height is 500 m.

5a. What is the plume rise, the effective plume height, the distance to maximum concentration, and the maximum concentration?

SOLUTION:

Executing SUPL indicates that the Pasquill Stability class is 1 (A). The buoyancy flux is 32.63. The unstable-neutral plume rise (using the '75 method, method 2 in WKBK2) is

83.6 m, giving an effective plume height of 113.6 m. The distance to maximum concentration is 0.465 km and the maximum concentration is 218 μg m^{-3}.

5b. At a distance twice the distance to the maximum concentration what is the concentration directly downwind and what fraction of the maximum concentration is this?

SOLUTION:

Using receptor method 2 to obtain concentrations downwind, 0.93 km was used to find that the groundlevel concentration at this distance is 62.6 μg m^{-3} or 28.7 percent of the maximum concentration.

5c. What is the crosswind distance to the 10 μg m^{-3} at these two downwind distances, 0.465 and 0.93 km?

SOLUTION:

The σ_y at 0.465 km is 106 m and at 0.93 km is 196 m. Using equation 2.11, the crosswind distance to the 1.0 x 10^{-5} g m^{-3} at 0.465 km is 0.263 km, and this distance at 0.93 km is 0.375 km. This can be checked using receptor method 2 and using the value of concentration for calculation of isopleths as 1.0 x 10^{-5} g m^{-3}.

5d. Verify that the concentration at a downwind distance of 0.93 km and a crosswind distance of 0.375 km is 10 μg m^{-3}.

SOLUTION:

Executing the program WKBK2 and using receptor methodology 3, with x = 0.93 and y = 0.375, gives the concentration at this position as 10.004 μg m^{-3}.

5e. If the wind direction is 5° different than was thought, what is the concentration at the position 0.93 km downwind that we previously thought was directly downwind?

SOLUTION:

Using program WKBK2 and receptor method 4, assume that the original receptor is at a particular azimuth (say 90°) from the source and at a distance range of 0.93 km. Then give the flow vector of the wind as 5 degrees different from the azimuth of the receptor (say 85° or 95°). The resulting concentration is 57.9 μg m^{-3} or 92 percent of that previously calculated.

Chapter 8.

5f. Use WKBK2 and receptor method 5 to verify the above by placing the source at coordinates (10.000, 10.000) and a receptor at (10.930, 10.000) and blowing the wind first toward an azimuth of 90°, and then toward azimuths of 85° and 95°. (Note: the scale factor used is 1.0, indicating that the coordinates are in km.)

SOLUTION:

Using a flow vector of 90° gives a concentration at the receptor of 62.6 μg m^{-3}. Using flow vectors of both 85° and 95° gives concentrations of 57.9 μg m^{-3}.

PROBLEM 6.

Consider the same source and time as in Problem 5. Assume that this source is operating at 70% capacity which causes the emissions to be 35 g s^{-1} and the exit velocity is 14 m s^{-1}. The exit temperature is 318 °K. What is the resulting plume rise, effective height of emission, distance to maximum concentration, and maximum groundlevel concentration?

SOLUTION:

The buoyancy flux for this condition of 70% operating capacity is 18.24. The unstable-neutral plume rise (using the '75 method, method 2 in WKBK2) is 54.0 m, giving an effective plume height of 84.0 m. Using A stability and the Pasquill-Gifford dispersion parameters to simulate dispersion over rural conditions, the distance to maximum concentration is 0.378 km and the maximum concentration is 243 μg m^{-3}. Note that because of the reduced plume rise for this reduced capacity, that in spite of the reduced emissions the maximum groundlevel concentration is about 11% higher than when the plant is operating at full capacity.

PROBLEM 7.

A small source with a stack 8 meters tall releases 1.5 g s^{-1} of a toxic material. Under stable conditions when the wind speed at stack top is 1.3 m s^{-1}, the plume rise from this stack is only 4 meters. A sonic anemometer mounted at 10 meters above ground on the installation indicates that the horizontal and vertical wind fluctuations over the hour with the above conditions are σ_a = 12.3 degrees and σ_e = 2.7 degrees. The shortest distance from this source to the property fenceline is 200 meters. The level-of-concern, LOC, from a health point of view for this material for one hour periods is 150 μg m^{-3}. For this hour do we need to be concerned about the emissions from this stack at or beyond the fenceline?

SOLUTION:

Using the program WKBK2 specifying the effective height as 12 meters and the fluctuation dispersion technique, receptor method 2, to calculate concentrations beneath the plume, show that at 0.2 km the concentration of 14.4 μg m^{-3} is well below the LOC.

However, by using the maximum concentrations receptor technique it is found that the maximum concentration under these conditions occurs at a distance downwind of 0.648 km and gives a maximum concentration of 174 µg m^{-3}, a level over the LOC. Returning to use the concentrations beneath the plume technique, and examining concentrations beyond the point of maximum it is found that under these conditions the concentration drops below 150 µg m^{-3} at a downwind distance of 1.015 km, just over 1 km from the source.

PROBLEM 8.

Related to Problem 7, what is the approximate maximum isopleth width of the LOC concentration isopleth? At the distance of this maximum isopleth width, what angle does this subtend looking from the source?

SOLUTION:

Using the program WKBK2 and the concentrations beneath the plume calculating the isopleth width, it is found that the maximum distance of the 1.5×10^{-4} concentration from the plume axis is 0.049 km or approximately 50 meters. This occurs for distances between 0.72 km and 0.76 km. The arctan of 50/750 is 3.8°. Since the distance to this concentration extends to both sides of the downwind axis, the angle as viewed from the source of the extent of the concentrations above the Level of Concern is 2 times 3.8 or 7.6°.

PROBLEM 9.

A combustion turbine is located in a rural area with a surface roughness, z_0, of approximately 0.2 m near 35.9° north latitude and 78.8° west longitude. The SO_2 emissions are 50 g s^{-1} from a 15-m-tall stack with a 4.5-m diameter. The exit velocity is 35 m s^{-1} and the exit temperature is 600 °K.

What is the mean plume height and the maximum concentration and at what distance does the maximum concentration occur on November 7, 1992 at 1500? The temperature is 68° F (293 °K), the cloud cover is 4/10, and the wind speed at 10 meters is 4 m s^{-1}. Wind changes and potential temperature with height are given in the following table. Assume the surface pressure is 1002 mb, the air temperature is 293 K, and the mixing height is 1200 m.

Height m	Wind Speed m s^{-1}	$\partial\theta/\partial z$ °K m^{-1}
15.	4.	-0.0001
100.	4.5	-0.0001
700.	6.	0.0001
1200.	6.	0.0150
1500.	5.5	

SOLUTION:

Using program SUPL for the above conditions indicates that the Pasquill stability class is C, the sensible heat flux is 61.1 watts m^{-2}, and the friction velocity is 0.435 m s^{-1}. Using program WKBK2 the buoyancy flux is calculated to be 889 m^4 s^{-3} and the Briggs '75 plume rise to be 569 m, giving an effective height of 584 m. Using the Pasquill-Gifford parameters gives the distance to the maximum concentration as 8.135 km and the maximum groundlevel concentration to be 5.2 µg m^{-3}.

Using the '91 plume rise, the top of the plume penetrates the mixing height, leaving 0.96 of the plume below the mixing height. The resulting plume rise is 795.6 m, giving an effective plume height of 810.6 m. The calculated distance to maximum concentration is 13.406 km and the maximum concentration is 2.8 µg m^{-3}.

PROBLEM 10.

At the site of the source in Problem 9, four hours later (1900) the cloud cover has increased to 6/10, and winds have become lighter, 3 m s^{-1} at 15 m. The wind and potential temperature structure are given in the following table.

Height m	Wind Speed m s^{-1}	$\partial\theta/\partial z$ °K m^{-1}
15.	3.	0.0150
100.	3.5	0.0100
700.	4.	0.0009
1200.	4.5	-0.0005
1500.	5.0	

What is the resulting distance to maximum concentration and the maximum concentration at this time?

SOLUTION:

SUPL indicates that the stability class is E, the heat flux is -4.7 watts m^{-2}, and the friction velocity is 0.270 m s^{-1}. WKBK2 gives 218.1 m for the Briggs '75 plume rise, giving an effective height of 233.1 m. The distance to maximum concentration is calculated to be 30.286 km and the maximum concentration to be 7.3 µg m^{-3}.

Using the '91 plume rise through layers, the plume rise is 226.7 m giving an effective height of 241.7. The distance to maximum concentration is 33.405 km and the maximum concentration is 6.4 µg m^{-3}.

PROBLEM 11.

What differences occur for the conditions of the previous problem in the estimates of distance to maximum concentration and maximum concentration if buoyancy-induced dispersion is considered?

SOLUTION:

Using buoyancy induced dispersion and '75 rise, the maximum groundlevel concentrations are increased from 7.2 to 10.0 μg m^{-3} and the distance to maximum is decreased from 30.286 to 19.432 km.

Using buoyancy induced dispersion and '91 rise through layers, the maximum groundlevel concentrations are increased from 6.4 to 9.1 μg m^{-3} and the distance to maximum is decreased from 33.405 to 20.000 km.

PROBLEM 12.

It is estimated that a burning dump emits 3 g s^{-1} of oxides of nitrogen. What is the concentration of oxides of nitrogen, averaged over approximately 10 minutes, from this source directly downwind at a distance of 3 km on an overcast night with a wind speed of 7 m s^{-1}? Assume this dump to be a groundlevel point source with no effective rise.

SOLUTION:

Overcast conditions with a wind speed of 7 m s^{-1} indicate that stability class D is most applicable. For x = 3 km and stability D, σ_y = 185 m, and σ_z = 65.1 m (from Table 2.5). Equation 2.5 for estimation of concentrations directly downwind (y = 0) from a groundlevel source is applicable:

$$\chi(x,0,0;0) \ = \ \frac{Q}{\pi \ u \ \sigma_y \ \sigma_z} \ = \ \frac{3}{\pi \ 7 \ 185 \ 65.1}$$

$$= \ 1.13 \times 10^{-5} \ g \ m^{-3} = \ 11.3 \ \mu g \ m^{-3} \text{ of oxides of nitrogen.}$$

PROBLEM 13.

It is estimated that 80 g s^{-1} of sulfur dioxide is being emitted from a petroleum refinery from an average effective height of 60 meters. At 0800 on an overcast winter morning with the surface wind 6 m s^{-1}, what is the groundlevel concentration directly downwind from the refinery at a distance of 500 meters?

SOLUTION:

For overcast conditions, D stability applies. With D stability at x = 500 m, σ_y = 36.1 m, σ_z = 18.3 m (from Table 2.5). Using equation 2.3:

$$\chi(x,0,0;H) = \frac{Q}{\pi \, u \, \sigma_y \, \sigma_z} \exp\left[-\frac{H^2}{2\,\sigma_z^2}\right]$$

$$\chi(500,0,0;60) = \frac{80}{\pi \; 6 \; 36.1 \; 18.3} \exp[-\,0.5\,(60/18.3)^2]$$

$$= \; 6.42 \times 10^{-3} \; \exp[-\,0.5\,(3.28)^2]$$

The exponential can be solved with a hand calculator or by using Table 2.1

$$= \; 6.42 \times 10^{-3} \; (4.61 \times 10^{-3})$$

$$= \; 29.6 \times 10^{-6} \; g \, m^{-3} \; = \; 29.6 \; \mu g \, m^{-3} \; of \; SO_2$$

PROBLEM 14.

Under the conditions of problem 13, what is the concentration at the same distance downwind but at a distance 50 meters from the x axis? That is: $\chi(500, 50, 0; 60) = ?$

SOLUTION:

Since we have determined the concentration directly downwind at groundlevel in the previous problem, we need to multiply that result by the exponential involving σ_y.

$$\chi(500,50,0;60) = \chi(500,0,0;60) \; \exp\left[-\frac{y^2}{2\,\sigma_y^2}\right]$$

$$= \; 29.6 \quad \exp[-\,0.5(50/36.1)^2]$$

$$= \; 29.6 \quad \exp[-\,0.5(1.39)^2] \; = \; 29.6 \; (0.381)$$

$$= \; 11.3 \; \mu g \, m^{-3} \; of \; SO_2$$

PROBLEM 15.

A power plant burns 10 tons per hour of coal containing 3 percent sulfur; the effluent is released from a single stack. On a sunny summer afternoon the wind at 10 meters above ground is 4 m s^{-1} from the northeast. The morning radiosonde taken at a nearby Weather Service station indicated that a frontal inversion aloft will limit the vertical mixing to 1500 meters. The effective height of emission is 150 meters. From Figure 2.5, what is the distance to the maximum groundlevel concentration and what is the concentration at this point?

SOLUTION:

To determine the emission rate, Q, the amount of sulfur burned is: 10 tons h^{-1} x 2000 lbs ton^{-1} x 0.03 sulfur = 600 lbs sulfur h^{-1}. Sulfur has a molecular weight of 32 and combines with O_2 with a molecular weight of 32; therefore for every mass unit of sulfur burned, there results two mass units of SO_2.

$$Q = (64/32) \text{ x } 600 \text{ lbs h}^{-1} \text{ x } 453.6 \text{ g lb}^{-1} = 544,320 \text{ g h}^{-1}$$

$$= 544,320/3600 \text{ s h}^{-1} = 151.2 \text{ g s}^{-1}$$

On a sunny summer afternoon the insolation should be strong. From Table 2.2, strong insolation and 4 m s^{-1} winds yields class B stability. From Figure 2.5 the distance to the point of maximum concentration is 1 km for an effective height of 150 meters and B stability. The $(\chi u/Q)_{max} = 7.5 \times 10^{-6} \text{ m}^{-2}$.

$$\chi_{max} = (\chi u/Q)_{max} \text{ } Q/u = 7.5 \times 10^{-6} \text{ x } 151.2/4$$

$$= 2.84 \times 10^{-4} \text{ g m}^{-3} = 284 \text{ } \mu\text{g m}^{-3} \text{ of } SO_2$$

PROBLEM 16.

For the power plant in problem 15, at what distance does the maximum groundlevel concentration occur and what is this concentration on an overcast day with wind speed 4 m s^{-1}?

SOLUTION:

On an overcast day the stability class would be D. From Figure 2.5 for D stability and an effective height of 150 m, the distance to the point of maximum groundlevel concentration is about 5.7 km and the $(\chi u/Q)_{max}$ is 3.0×10^{-6}.

$$\chi_{max} = (\chi u/Q)_{max} \text{ x } Q/u = 3.0 \times 10^{-6} \text{ x } 151.2/4 = 113 \text{ } \mu\text{g m}^{-3}$$

This can be verified by application of the program WKBK2. The program gives the distance to the max as 5.617 km and the maximum concentration as 112 μg m^{-3}.

PROBLEM 17.

For the conditions given in problem 15, draw a graph of groundlevel centerline sulfur dioxide concentration with distance from 100 meters to 20 km. Use log-log graph paper.

SOLUTION:

Using the program WKBK2 and receptor method 2, the concentrations are determined for a number of downwind distances from 0.3 km to 100 km using the conditions for problem 15. These concentrations are given in the following table. Also given are the concentrations with an unlimited mixing height.

Downwind Distance km	σ_y m	σ_z m	χ $\mu g\ m^{-3}$ with 1200 mixing height	χ $\mu g\ m^{-3}$ with unlimited mixing height
0.362	61.9	36.3	1.03×10^{-6}	1.03×10^{-6}
0.4	67.7	40.0	3.93×10^{-6}	3.93×10^{-6}
0.5	82.8	51.1	3.82×10^{-5}	3.82×10^{-5}
0.6	97.5	62.4	1.10×10^{-4}	1.10×10^{-4}
0.7	112.	73.9	1.85×10^{-4}	1.85×10^{-4}
0.8	126.	85.6	2.40×10^{-4}	2.40×10^{-4}
0.9	140.	97.4	2.69×10^{-4}	2.69×10^{-4}
1.0	154.	109.	2.79×10^{-4}	2.79×10^{-4}
1.1	168.	121.	2.75×10^{-4}	2.75×10^{-4}
1.2	181.	134.	2.64×10^{-4}	2.64×10^{-4}
1.3	195.	146.	2.50×10^{-4}	2.50×10^{-4}
1.4	208.	158.	2.33×10^{-4}	2.33×10^{-4}
1.5	221.	171.	2.17×10^{-4}	2.17×10^{-4}
1.6	234.	183.	2.00×10^{-4}	2.00×10^{-4}
1.7	247.	196.	1.85×10^{-4}	1.85×10^{-4}
1.8	260.	208.	1.71×10^{-4}	1.71×10^{-4}
1.9	273.	221.	1.58×10^{-4}	1.58×10^{-4}
2.0	286.	234.	1.47×10^{-4}	1.47×10^{-4}
2.5	348.	299.	1.02×10^{-4}	1.02×10^{-4}
3.	409.	365.	7.41×10^{-5}	7.41×10^{-5}
4.	527.	500.	4.36×10^{-5}	4.36×10^{-5}
5.	641.	639.	2.86×10^{-5}	2.86×10^{-5}
6.	753.	780.	2.01×10^{-5}	2.01×10^{-5}
7.	861.	924.	1.51×10^{-5}	1.49×10^{-5}
8.	967.	1070.	1.20×10^{-5}	1.15×10^{-5}
9.	1071.	1218.	1.01×10^{-5}	9.15×10^{-5}
10.	1174.	1367.	8.83×10^{-6}	7.45×10^{-6}
15.	1666.	2133.	6.03×10^{-6}	3.38×10^{-6}
20.	2133.	2924.	4.71×10^{-6}	1.94×10^{-6}
30.	3011.	4562.	3.34×10^{-6}	1.03×10^{-6}
50.	4627.	5000.	2.17×10^{-6}	6.61×10^{-7}
70.	6120.	5000.	1.64×10^{-6}	5.00×10^{-7}
100.	8201.	5000.	1.23×10^{-6}	3.73×10^{-7}

These concentrations are plotted in Figure 8.1.

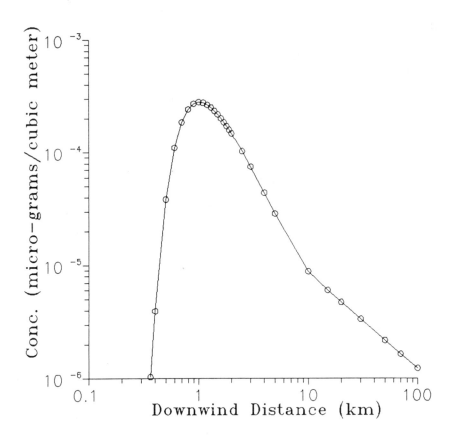

Figure 8.1 Concentration as a function of downwind distance (Problem 17).

Note that the 1200-m mixing height doesn't affect the groundlevel concentrations until a downwind distance of 10 km is reached. Note also that the σ_z is not allowed to grow after reaching 5000 m. If the plume has expanded this much in the vertical, the top is reaching the bottom of the stratosphere which is usually very stable. In almost all situations there will be a lower mixing height that will limit the vertical spreading of the plume.

PROBLEM 18.

For the conditions given in problem 15, draw a graph of groundlevel concentration versus crosswind distance at a downwind distance of 1 km.

Chapter 8.

SOLUTION:

From problem 15 the groundlevel centerline concentration at 1 km is 284 μg m^{-3}. To determine the concentrations at distances y from the x-axis, the groundlevel centerline concentration must be multiplied by the factor: $\exp[-0.5(y/\sigma_y)^2]$. The value of σ_y at 1 km is 154 m. Values for these computations are given in following table.

y, m	y/σ_y	$\exp[-0.5(y/\sigma_y)^2]$	$\chi(1$ km,y,0$)$
± 0	0.	1.00	284
± 25	0.162	0.987	275
± 50	0.324	0.949	264
± 75	0.487	0.888	247
±100	0.649	0.810	226
±125	0.811	0.720	200
±150	0.973	0.623	173
±175	1.14	0.525	146
±200	1.30	0.431	120
±225	1.46	0.345	96.0
±250	1.62	0.268	74.7
±275	1.78	0.204	56.7
±300	1.95	0.150	41.9
±325	2.11	0.108	30.1
±350	2.27	0.0759	21.1
±375	2.43	0.0518	14.4
±400	2.60	0.0345	9.60
±425	2.76	0.0223	6.22
±450	2.92	0.0141	3.92
±475	3.08	8.66×10^{-3}	2.41
±500	3.24	5.18×10^{-3}	1.44

These concentrations are plotted in Figure 8.2.

PROBLEM 19.

For the conditions given in problem 15, determine the position of the 10^{-5} g m^{-3} groundlevel isopleth, and determine its area.

SOLUTION:

From the solution to Problem 17, the graph shows that the 10^{-5} g m^{-3} isopleth intersects the x-axis at approximately 400 meters and 11 km downwind. The following table gives the distances from the beneath-the-plume axis to the 10^{-5} isopleth. These are actually isopleth half-widths.

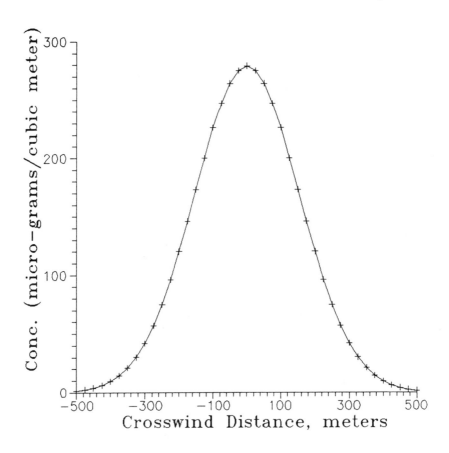

Figure 8.2 Concentration as a function of crosswind distance (Problem 18).

The area of an ellipse is π a b where a is the semi-major axis and b is the semi-major axis. The semi-major axis is $(11000 - 400)/2 = 5300$ m. The semi-minor axis is 930 m. The area is 15.5×10^6 m^2 or 15.5 km^2.

PROBLEM 20.

For the conditions given in problem 15, determine the profile of concentration with height from groundlevel to $z = 450$ meters at $x = 1$ km, $y = 0$ meters, and draw a graph of concentration against height above ground.

SOLUTION:

Equation 2.1 is used to determine concentrations at various heights above ground. At 1 km, from Table 2.5, $\sigma_y = 154$ m, $\sigma_z = 109$ m.

x, km	σ_y, m	χ(centerline) $\mu g\ m^{-3}$	Isopleth half-width, y m
0.42	70.7	7.2	0.0
0.44	73.8	12.2	47.
0.46	76.8	19.0	87.
0.48	80.0	27.7	114.
0.5	82.8	38.2	136
0.6	97.5	110	214
0.7	112	185	271
0.8	126	240	318
0.9	140	269	360
1.0	154	279	398
1.5	221	217	549
2.0	286	146	662
3.0	409	74.1	819
4.0	527	43.6	905
5.	641	28.6	930 *
6.	753	20.5	903
7.	861	16.0	837
8.	967	13.5	746
9.	1071	11.9	626
10.	1174	10.7	443
10.2	1194	10.5	389
10.4	1215	10.4	325
10.6	1235	10.2	240
10.8	1255	10.0	85
10.85	1260	9.98	0

*   Maximum isopleth width

$$\frac{Q}{2\ \pi\ u\ \sigma_y\ \sigma_z} = \frac{151.2}{2\ \pi\ 4\ 154\ 109} = 3.58 \times 10^{-4}\ g\ m^{-3}$$

Values for the estimation of χ(z) are given in the following table.

This concentration variation with height is shown in Figure 8.3.

PROBLEM 21.

For the conditions given in problem 15, determine the distance at which the groundlevel centerline concentration equals the plume centerline concentration at 150 meters above ground. Verify by computation at χ(x,0,0) and χ(x,0,150).

Downwind Distance, km	z, m	$\chi(z)$ $\mu g\ m^{-3}$
1.	0.	279.
1.	30.	288.
1.	60.	311.
1.	90.	339.
1.	120.	361.
1.	150.	365.
1.	180.	348.
1.	210.	309.
1.	240.	255.
1.	270.	196.
1.	300.	139.
1.	330.	92.0
1.	360.	56.4
1.	390.	32.1
1.	420.	16.9
1.	450.	8.26

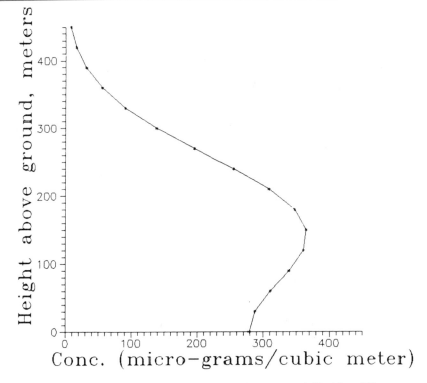

Figure 8.3 Concentration as a function of height above ground (Problem 20).

SOLUTION:

The distance at which concentrations at the ground and at plume height are equal should occur where $\sigma_z = 0.91\ H$ according to the discussion in Section 4.5. For B stability and $H = 150$ m, $\sigma_z = 0.91\ (150) = 136$ m occurs at an x of approximately 1.22 km. At this distance σ_y is approximately 184 m.

Verifying:

$$\chi(1.22,0,0;150) = \frac{Q}{\pi\ u\ \sigma_y\ \sigma_z}\ \exp[-0.5(H/\sigma_z)^2]$$

$$= \frac{151.2}{\pi\ 4\ 184\ 136}\ \exp[-0.5(150/136)^2]$$

$$= 4.81 \times 10^{-4}\ \exp[-0.5(1.10)^2]$$

$$= 4.81 \times 10^{-4}\ (0.544) = 262\ \mu g\ m^{-3}$$

$$\chi(1.22,0,150;150) = \frac{Q}{2\ \pi\ u\ \sigma_y\ \sigma_z}\ \{\exp[-0.5(z-H/\sigma_z)^2] + \exp[-0.5(z+H/\sigma_z)^2]\}$$

$$= \frac{151.2}{2\ \pi\ 4\ 184\ 136}\ \{\exp[-0.5(0/136)^2] + \exp[-0.5(300/136)^2]\}$$

$$= 2.40 \times 10^{-4}\ \{1. + \exp[-0.5(2.21)^2]\} = 2.40 \times 10^{-4}\ \{1. + 0.0878\}$$

$$= 2.40 \times 10^{-4}\ \{1.0878\} = 262\ \mu g\ m^{-3}$$

The two concentrations are the same.

PROBLEM 22.

For the power plant in problem 15, what will the maximum groundlevel concentration be beneath the plume centerline and at what distance will it occur on a clear night with wind speed 4 m s^{-1}? Assume that the effective height is at 150 m. (In an actual situation the stable effective height would probably be lower.)

SOLUTION:

A clear night with wind speed 4 m s^{-1} indicates E stability class. From Figure 2.5, the maximum concentration occurs at about 13 km, and the maximum $\chi u/Q$ is 1.7×10^{-6}.

$$\chi = (\chi u/Q) \times (Q/u) = 1.7 \times 10^{-6} \times (151.2/4) = 6.4 \times 10^{-5} \text{ g m}^{-3}$$

$$= 64 \text{ } \mu g \text{ m}^{-3} \text{ of } SO_2$$

PROBLEM 23.

For the situation in problem 22, what would the fumigation concentration be the next morning at this point (x = 13 km) when superadiabatic lapse rates extend to include most of the plume and it is assumed that wind speed and direction remain unchanged?

SOLUTION:

The concentration during fumigation conditions is given by Equation 4.8 with the exponential involving y equal to 1.

$$\chi_F(x,0,0;H) = Q/[(2\pi)^{0.5} u \sigma_{yF} h_i]$$

For the stable conditions which were assumed to be stability class E, from Table 2.5 at x = 13 km, $\sigma_y = 514$ m and $\sigma_z = 89.4$ m. Using $h_i = H + 2 \sigma_z$ to solve for h_i:

$$h_i = H + 2 \sigma_z = 150 + 2 (89.4) = 328.8 \text{ m}$$

From the horizontal spreading suggested by Equation 4.9:

$$\sigma_{yF} = 514 + (150/8) = 533 \text{ m.}$$

$$\chi_F(x,0,0;H) = Q/[(2\pi)^{0.5} u \sigma_{yF} h_i] = 151.2/[2.5066 \times 4 \times 533 \times 328.8]$$

$$= 8.6 \times 10^{-5} \text{ g m}^{-3} = 86 \text{ } \mu g \text{ m}^{-3}$$

Note that the fumigation concentrations under these fumigation conditions are about 1.3 times the maximum groundlevel concentrations that occurred during the stable conditions during the night in the previous problem.

PROBLEM 24.

An air sampling station is located at an azimuth of 203° from a cement plant at a distance of 1500 meters. The cement plant releases fine particulates (less than 10 μm diameter) at the rate of 75 pounds per hour from an effective height of 30 meters when the wind is

3 m s^{-1}. What is the contribution from the cement plant to the total suspended particulate concentration at the sampling station when the wind is from 30° at 3 m s^{-1} on a clear day in the late fall at 1600?

SOLUTION:

For this season and time of day the C stability should apply. Since the sampling station is off the plume axis, the x and y distances can be calculated:

$$x = 1500 \cos 7° = 1489 \text{ m}$$

$$y = 1500 \sin 7° = 183 \text{ m}$$

The source strength is:

$$Q = 75 \text{ lb hr}^{-1} \times 0.126 \, [(\text{g s}^{-1})/(\text{lb hr}^{-1})] = 9.45 \text{ g s}^{-1}$$

At this distance, 1489 m, for stability C, $\sigma_y = 149$ m, $\sigma_z = 88.6$ m. The contribution of this plant to the concentration at this monitor can be calculated from Equation 2.2:

$$\chi(x,y,0;H) = \frac{Q}{\pi \, u \, \sigma_y \, \sigma_z} \exp\left[-\frac{y^2}{2\sigma_y^2}\right] \exp\left[-\frac{H^2}{2\sigma_z^2}\right]$$

$$= \frac{9.45}{\pi \, 3 \, 149 \, 88.6} \exp[-0.5(183/149)^2] \, \exp[-0.5(30/88.6)^2])$$

$$= 7.595 \times 10^{-5} \quad \exp(-7.54) \quad \exp(-0.0573)$$

$$= 7.595 \times 10^{-5} \quad 0.470 \quad 0.944$$

$$= 3.37 \times 10^{-5} \text{ g m}^{-3} = 33.7 \text{ } \mu\text{g m}^{-3}$$

PROBLEM 25.

A proposed source is to emit 72 g s^{-1} of SO_2 from a stack 30 meters high with a diameter of 1.5 meters. The effluent gases are emitted at a temperature of 250°F (394°K) with an exit velocity of 13 m s^{-1}. Plot on log-log paper a graph of maximum groundlevel concentration as a function of wind speed for stability classes B and D. Determine the critical wind speed for these stabilities, i.e., the wind speed that results in the highest concentrations. Assume that the design atmospheric temperature is 20°C (293°K).

SOLUTION:

Using the information above, the buoyancy flux is found to be 18.38 and the product of ΔH times u is 190.2. Values of maximum concentration as a function of wind speed are given in the following table:

Stability Class	Wind Speed	ΔH	H	χ_{max} $\mu g\ m^{-3}$	
B	0.5	380.4	410.4	171.6	
B	1.0	190.2	220.2	264.6	
B	1.5	126.8	156.8	326.6	
B	2.	95.1	125.1	369.1	
B	3.	63.4	93.4	418.5	
B	5.	38.0	68.0	446.8	*
B	7.	27.2	57.2	438.2	
D	0.5	380.4	410.4	33.5	
D	1.0	190.2	220.2	82.5	
D	1.5	126.8	156.8	127.9	
D	2.	95.1	125.1	168.1	
D	3.	63.4	93.4	230.3	
D	5.	38.0	68.0	296.2	
D	7.	27.2	57.2	321.9	
D	10.	19.0	48.4	335.8	*
D	15.	12.7	40.8	323.4	
D	20.	9.5	37.0	299.0	

*   Maximum for this stability class

This information is graphed in Figure 8.4.

PROBLEM 26.

A proposed pulp processing plant is expected to emit 1/2 ton per day of hydrogen sulfide from a single stack. The company property extends a minimum of 1500 meters from the proposed location. The nearest receptor is a small town of 500 inhabitants 1700 meters northeast of the plant. Plant managers have decided that it is desirable to maintain concentrations below 20 ppb (parts per billion by volume), or approximately 2.9×10^{-5} g m^{-3}, for any period greater than 30 minutes. Wind direction frequencies indicate that winds blow from the proposed location toward this town between 10 and 15 percent of the time. What height stack should be erected? It is assumed that a design wind speed of 2 m s^{-1} will be sufficient, since the effective stack rise will be quite great with winds less than 2 m s^{-1}. Other than this stipulation, assume that the physical stack height and effective stack height are the same, to incorporate a safety factor.

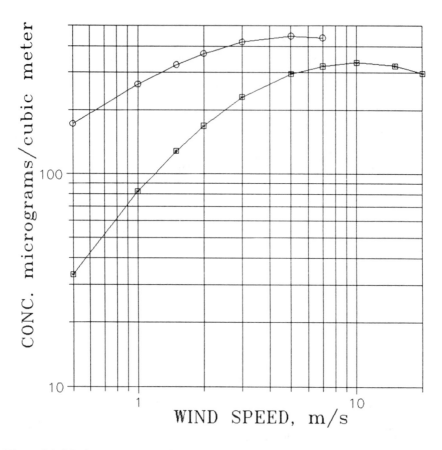

Figure 8.4 Maximum concentration as a function of wind speed (Problem 25).

SOLUTION:

The source strength is: $Q = \dfrac{1000 \text{ lbs day}^{-1} \times 453.6 \text{ g lb}^{-1}}{86400 \text{ s day}^{-1}} = 5.25 \text{ g s}^{-1}$

Starting with Equation 2.6 it is found that if the release were from groundlevel, the product of σ_y times σ_z would need to be $5.25/(\pi\ 2\ 2.9 \times 10^{-5}) = 28813.\ \text{m}^2$. It is seen from Table 2.5 that at 1500 meters, this value of 2.88×10^4 is exceeded for A and B stabilities, indicating that for those stabilities the concentration will drop below the desired level even if the release is from groundlevel. However for C stability the product of σ_y times σ_z is 1.32×10^4. This is 0.458 of the value needed. Therefore an elevated height of release is needed that will cause $\exp[-0.5(H/\sigma_z)^2]$ to be 0.458. From Table 2.1 it is seen that if the ratio of H/σ_z is 1.25, the exponential will be the correct value. Since for class C stability σ_z is 88.6 at a downwind distance of 1500 m (from Table 2.5), H must be 111 m for this to be true. Using the program WKBK2 with receptors beneath the plume and using x as 1.5 km, for an effective height of 111 m with C stability the concentration is 2.89 x 10^{-5}. The program is also used to verify that with this height of release, the concentrations for D, E, and F stabilities are well below the desired concentration.

PROBLEM 27.

In problem 26 assume that the stack diameter is to be 1.5 m, the temperature of the effluent 350 K, and the stack gas velocity 13.7 m s^{-1}. From the appropriate equation for effective stack height and the effective height determined in problem 26, determine the physical stack height required to satisfy the conditions in problem 26. In estimating plume rise assume the ambient air temperature is 68°F (20°C).

SOLUTION:

From the specified conditions, the buoyancy flux is 12.3. For a wind speed of 2 m s^{-1} and unstable conditions which would correspond to C stability, the plume rise is calculated to be 70.4 m. Subtracting this from 111 m, the required stack height is 40.6 m. Using the WKBK2 program to check for C stability and 2 m s^{-1}, the calculated concentration at 1.5 km is 28.9 μg m^{-3}. Checking for wind speeds of 3, 4, 5, and 7 yields lower concentrations.

PROBLEM 28.

A dispersion study is being made over relatively open terrain with fluorescent particles whose size yields 1.8 x 10^{10} particles per gram of tracer. Sampling is by membrane filters through which 9 x 10^{-3} m^3 of air is drawn each minute. A study involving a 1-hour release, which can be considered from groundlevel, is to take place during conditions forecast to be slightly unstable with winds 5 m s^{-1}. It is desirable to obtain a particle count of at least 20 particles upon membrane filters located at groundlevel 2 km from the plume centerline on the sampling arc 8 km from the source. What should the total release be, in grams, for this run.

SOLUTION:

The total dosage at the sampler is determined by the desired sample in grams divided by the sampling rate:

$$D_T = \frac{20 \text{ particles}}{1.8 \times 10^{10} \text{ particles g}^{-1}} \cdot \frac{60 \text{ s min}^{-1}}{9 \times 10^{-3} \text{ m}^3 \text{ min}^{-1}}$$

$$= 1200/16.2 \times 10^7 = 7.41 \times 10^{-6} \text{ g s m}^{-3}$$

The total dosage is given in g s m^{-3} from equation 4.15:

$$D_T(x,y,0;H) = \frac{Q_T}{\pi u \sigma_y \sigma_z} \exp\left[-\frac{y^2}{2\sigma_y^2}\right]$$

where Q_T is the total release in grams. For slightly unstable conditions, that is, C stability, at x = 8 km, σ_y = 672 m, and σ_z = 410 m. Solving for Q_T yields 2689. g.

No correction has been made for the fact that the release is for an hour and the σ_y is for a time period less than an hour (according to Pasquill (1976) the σ_y is for 3 minutes).

PROBLEM 29.

A release of 3 kg of fluorescent particles is made based on the results of the computation in problem 28. For the conditions above, the stability class is C stability and wind speed 5 m s^{-1}. The crosswind-integrated groundlevel dosage along the 8-km arc is determined from the samplers along this arc to be 8.2 x 10^{-1} g s m^{-2}. What is the effective σ_z for this run?

SOLUTION:

The crosswind integrated dosage is given by:

$$D_{CWI} = \frac{2 Q_T}{(2\pi)^{0.5} u \sigma_z}$$

$$\sigma_z = (2 \times 3000)/(2.5066 \; 5 \; 0.82) = 583.8 \text{ m}$$

PROBLEM 30.

At a point directly downwind from a groundlevel source, the one-hour concentration is estimated to be 58 µg m^{-3}. What would you expect the peak three-minute concentration to be during this one-hour period?

SOLUTION:

According to the work of Hino shown in Table 4.3, the three-minute peak concentration can be expected to be four times the one-hour concentration. Therefore the peak three-minute concentration during this hour would be expected to be 232 µg m^{-3}.

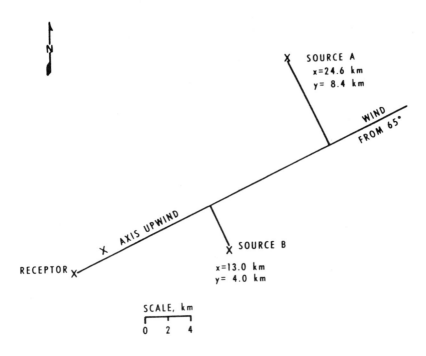

Figure 8.5 Locations of sources and receptor (Problem 31).

PROBLEM 31.

Two sources of SO_2 are shown as points A and B in Figure 8.5. On a sunny summer afternoon the surface wind is from 60° at 6 m s^{-1}. Source A is a power plant emitting 1450 g s^{-1} SO_2 from two stacks whose physical height is 120 meters and whose ΔH, for unstable-neutral conditions is ΔH (m) = 624 (m^2 s^{-1})/u (m s^{-1}). Source B is a refinery emitting 126 g s^{-1} SO_2 from an effective height of 60 meters. The wind measured at 160 meters on a nearby TV tower is from 70° at 8.5 m s^{-1}. Assuming that the mean direction of travel of both plumes is 245°, and there are no other sources of SO_2, what is the concentration of SO_2 at the receptor shown in the figure?

SOLUTION:

For a sunny summer afternoon with wind speed 6 m s^{-1}, the stability class is expected to be Pasquill Class C. Equation 2.2 is to be used.

Chapter 8.

For source A, $\Delta H = 624/8.5 = 73.4$ m, $H = 193.4$ m, $x = 24.6$ km, $y = 8.4$ km. At $x = 24.6$ km and C stability, σ_y is approximately 1815 m from Table 2.5 and σ_z is approximately 1145 m. The concentration due to source A, χ_A, is:

$$\chi_A = [1450/(\pi\ 8.5\ 1815\ 1145)]\ \exp[-0.5(8400/1815)^2]\ \exp[-0.5(193.4/1145)^2]$$

$$= 2.613 \times 10^{-5} \quad 2.233 \times 10^{-5} \quad 0.9858$$

$$= 5.75 \times 10^{-10}\ \text{g m}^{-3}$$

For source B, $H = 60$ m, $u = 7$ m s^{-1}, $x = 13$ km, $y = 4$ km, At $x = 13$ km and C stability, σ_y is 1035 m and σ_z is 639 m. The concentration due to source B, χ_B, is:

$$\chi_B = [126/(\pi\ 7\ 1035\ 639)]\ \exp[-0.5(4000/1035)^2]\ \exp[-0.5(60/639)^2]$$

$$= 8.663 \times 10^{-6} \quad 5.710 \times 10^{-4} \quad 0.9956$$

$$= 4.925 \times 10^{-9}\ \text{g m}^{-3}$$

The total concentration, χ, is the sum of χ_A and χ_B.

$$\chi = 5.75 \times 10^{-10} + 4.925 \times 10^{-9}$$

$$= 5.5 \times 10^{-9}\ \text{g m}^{-3}$$

PROBLEM 32.

A stack 15 meters high emits 3 g s^{-1} of an air pollutant. The surrounding terrain is relatively flat except for a rounded hill about 3 km to the northeast whose crest extends 15 meters above the stack top. What is the highest 3- to 15-minute concentration of this pollutant that can be expected on the facing slope of the hill on a clear night when the wind is blowing directly from the stack toward the hill at 4 m s^{-1}? Assume that ΔH is less than 15 m. How much does the wind have to shift so that concentrations at this point drop below 10^{-7} g m^{-3}?

SOLUTION:

A clear night with wind 4 m s^{-1} indicates Pasquill stability class E. Equation 2.4 for the concentration at the plume centerline can be used to see what the concentrations would be in the elevated plume at a distance of 3 km. Under E stability at 3 km, $\sigma_y = 138$ m, $\sigma_z = 42.2$.

$$\chi = [3/(2\ \pi\ 138\ 42.2]\ \{1. + \exp[-0.5(30/42.2)^2]\}$$

$$= 8.199 \times 10^{-5} \quad 1.777$$

$$= 1.46 \times 10^{-4}\ \text{g m}^{-3}$$

No eddy reflection is included from the hill surface. It is felt that the concentrations on the hill surface are not likely to be any higher than the plume centerline concentration at this distance over flat terrain.

To determine the crosswind distance from the plume centerline to produce a concentration of 10^{-7} g m^{-3}, Equation 2.11 is used:

$$y = [2 \ln(1.46 \times 10^{-4}/10^{-7})]^{0.5} \, \sigma_y$$

$$= 3.817 \quad 138$$

$$= 527 \text{ m}$$

$$\tan a = 527/3000 = 0.176$$

$$a = 10°$$

A wind shift of 10° is required to reduce the concentration to 10^{-7} g m^{-3}.

PROBLEM 33.

An inventory of SO_2 emissions has been conducted in an urban area by square areas, 5000 ft (1524 meters) on a side. The emissions from one such area are estimated to be 6 g s^{-1} for the entire area. This square is composed of residences and a few small commercial establishments. What is the concentration resulting from this area at the center of the adjacent square to the north when the wind is blowing from the south on a thinly overcast night with the wind at 2.5 m s^{-1}? The average effective stack height of these sources is assumed to be 20 meters.

SOLUTION:

A thinly overcast night with wind speed 2.5 m s^{-1} indicates stability of class E. (It may actually be more unstable, since this is in a built-up area.) To allow for the area source, let $\sigma_{yo} = 1524/4.3 = 354.$ m. Using the dispersion parameters for urban conditions given in Table 4.5, $\sigma_y = 354$ for E-F stability at a distance of approximately 6. km. This is the virtual x_y distance. For x = 1524m, $\sigma_z = 67$ m. At x + x_y = 7524 m, $\sigma_y = 415$ m. The concentration at the center of the adjacent area source downwind is:

$$\chi = [Q/(\pi \, u \, \sigma_y \, \sigma_z)] \exp [-0.5(H/\sigma_z)^2]$$

$$= [6/(\pi \; 2.5 \; 415 \; 67)] \exp [-0.5(20/67)^2]$$

$$= 2.75 \times 10^{-5} \; 0.956 = 2.63 \times 10^{-5} \text{ g m}^{-3} = 26.3 \; \mu\text{g m}^{-3}$$

PROBLEM 34.

An estimate is required of the total hydrocarbon concentration 300 meters downwind of an expressway at 1730 on an overcast day with wind speed 4 m s^{-1}. The expressway runs north-south and the wind is from the west. The measured traffic flow is 8000 vehicles per hour during this rush hour, and the average speed of the vehicles is 40 miles per hour. At this speed, the average vehicle is expected to emit 2 x 10^{-2} g s^{-1} of total hydrocarbons.

SOLUTION:

The expressway may be considered as a continuous infinite line source. To obtain an emission rate, q, in g m^{-1} s^{-1}, the number of vehicles per meter of highway must be calculated and multiplied by the emission per vehicle.

$$\text{Vehicles/meter} = \text{Flow veh hr}^{-1}/(\text{Average speed mi hr}^{-1} \ 1609 \text{ m mi}^{-1})$$

$$= 8000/(40 \ 1609) \qquad = 0.124 \text{ veh m}^{-1}$$

$$q = 0.124 \text{ veh m}^{-1} \ 0.02 \text{ g s}^{-1} \text{ veh}^{-1} \quad = 2.5 \times 10^{-3} \text{ g s}^{-1} \text{ m}^{-1}$$

Under overcast conditions with wind speed 4 m s^{-1}, Pasquill stability class D applies. Under D, at x = 300 m, σ_z = 12.1 m for the rural Pasquill-Gifford parameters. Assuming that H = 0, Equation 4.21 yields:

$$\chi(300,0,0;0) = 2 q/[(2 \pi)^{0.5} \sigma_z u] = 2 \ 2.5 \times 10^{-3}/[2.5066 \ 12.1 \ 4]$$

$$= 4.12 \times 10^{-5} \text{ g m}^{-3} = 41.2 \ \mu g \text{ m}^{-3} \text{ of total hydrocarbons}$$

If we assume that these are urban conditions, at 300 m, σ_z = 40.2. This would cause the concentration to be 12.1/40.2 of that previously calculated, that is, 12.4 μg m^{-3}.

PROBLEM 35.

A line of burning agricultural waste can be considered a finite line source 150 m long. It is estimated that the total emission of organics is at a rate of 90 g s^{-1}. What is the 3- to 15-minute concentration of organics at a distance of 400 m directly downwind from the center of the line when the wind is blowing at 3 m s^{-1} perpendicular to the line? Assume that it is 1600 on a sunny fall afternoon. Also, what is the concentration directly downwind from one end of the source?

SOLUTION:

Late afternoon at this time of year implies slight insolation, which with 3 m s^{-1} winds yields Pasquill stability class C. For C stability at x = 400 m, σ_y = 44.6 m, σ_z = 26.4 m. Equation 4.23 is appropriate.

$$q = Q/150 = 90/150 = 0.6 \text{ g s}^{-1} \text{ m}^{-1}$$

For a point downwind of the center of the line,

$$p_1 = -75/\sigma_y = -75/44.6 = -1.68; \qquad p_2 = +1.68$$

The area under the Gaussian curve between s of -1.68 and +1.68 can be interpolated from Table 4.1 as $0.95338 - 0.04662 = 0.90676$

$$\chi(x,0,0;0) = 2q/[(2\pi)^{0.5}\sigma_z u] \quad \text{(Area under the curve between the limits)}$$

$$\chi(400,0,0;0) = 2 \times 0.6/[2.5066 \ 26.4 \ 3](0.90676)$$

$$= 6.04 \times 10^{-3} \ (0.90676) = 5.48 \times 10^{-3} \text{ g m}^{-3}$$

For a point downwind of one of the ends of the line:

$$p_1 = 0 \qquad p_2 = +150/44.6 = +3.36$$

The area under the Gaussian curve between s = 0 and s = +3.36 is $0.99960 - 0.5 = 0.49960$

$$\chi(400,75,0;0) = 6.04 \times 10^{-3} \ (0.49960) = 3.02 \times 10^{-3} \text{ g m}^{-3}$$

PROBLEM 36.

A core melt-down of a power reactor that has been operating for over a year occurs at 0200, releasing 1.5×10^6 curies of activity (1 second after the accident) into the atmosphere of the containment vessel. This total activity can be expected to decay according to $(t/t_o)^{-0.2}$. It is estimated that about 5.3×10^4 curies of this activity is due to iodine-131, which has a half-life of 8.04 days. The reactor building is hemispherically shaped with a radius of 20 meters. Assume the leak rate of the building is 0.1 percent per day. The accident has occurred on a relatively clear night with wind speed 2.5 m s^{-1}. What is the concentration in the air 3 km directly downwind from the source at 0400 due to all radioactive material? Due to iodine-131?

SOLUTION:

Emission rate = leak rate x activity (corrected for decay).

Leak rate = $0.001 \text{ day}^{-1}/86400 \text{ s day}^{-1} = 1.157 \times 10^{-8} \text{ s}^{-1}$

Emission rate of all products at 0400, 7200 s after 0200:

$$Q_A = 1.157 \times 10^{-8} \quad 1.5 \times 10^6 \ (t/t_o)^{-0.2} = 1.74 \times 10^{-2} \ (7200/1)^{-0.2}$$

$$= 1.74 \times 10^{-2} \ (0.169) = 2.94 \times 10^{-3} \text{ curies s}^{-1}$$

For a clear night with wind speed 2.5 m s^{-1}, Pasquill stability class F applies. Approximate the initial size of the cloud by 2.15 σ_{yo} = 2.15 σ_{zo} = the radius of the containment shell = 20 m. This gives σ_{yo} = σ_{zo} = 9.3 m. The virtual distances to account for this initial size are:

$$x_y = 0.24 \text{ km}, \quad x_z = 0.55 \text{ km}$$

$$\text{At } x = 3000 \text{ m}, \quad x + x_y = 3240 \text{ m}, \quad \sigma_y = 98.5 \text{ m}$$
$$x + x_z = 3550 \text{ m}, \quad \sigma_z = 29.2 \text{ m}$$

$$\chi(x,0,0;0) = Q/(\pi\, u\, \sigma_y\, \sigma_z) = Q/(\pi\ 2.5\ 98.5\ 29.2)$$

$$= 4.43 \times 10^{-5}\ Q$$

For the concentration at 0400, 3000 m downwind due to all radioactivity,

$$\chi_A = 4.43 \times 10^{-5}\ 2.94 \times 10^{-3} = 1.30 \times 10^{-7} \text{ curies m}^{-3}$$

To determine decay of materials with the half-life given, multiply by exp(-0.693 t/L) where L is half-life. The half-life of iodine^{131} is 8.04 days = 6.95 x 10^5 seconds.

The source strength of I^{131}:

$$Q_I = \text{leak rate times initial iodine times decay}$$

$$= 1.157 \times 10^{-8}\quad 5.3 \times 10^4 \quad \exp[-0.693\ 7200/(6.95 \times 10^5)]$$

$$= 6.13 \times 10^{-4}\ (0.993) = 6.09 \times 10^{-4} \text{ curies s}^{-1}$$

The concentration at 0400, 3000 m downwind due to I^{131},

$$\chi_I = 4.43 \times 10^{-5}\ 6.09 \times 10^{-4} = 2.70 \times 10^{-8} \text{ curies m}^{-3}$$

PROBLEM 37.

Five kg of a toxic substance was accidently released yesterday evening at 2000 (an hour after sunset). The release can be considered instantaneous and at groundlevel. The character of the surrounding neighborhood can be considered rural. This substance is considered to have a health-related level-of-concern of 1250 µg m^{-3} for 30-min exposures. The nearest distance to the property line from the accident location is 400 m. At the time this occurred there was 2/10 cloud and the 10-meter wind was 4 m s^{-1}. Were concentrations exceeding the LOC (level-of-concern) likely to have occurred off property and if they did to what distance do we need to be concerned? Assume that $\sigma_x = \sigma_y$. In determining a transport wind speed assume that a wind for a height of 2 m above ground would be appropriate. Determine this from the power-law wind profile equation in order to extrapolate downward from the wind speed at 10 meters height.

SOLUTION:

From Table 2.2 at night with a 10-meter wind speed of 4 m s^{-1}, the Pasquill stability class would be E. The power-law wind profile exponent for this stability for rural conditions is 0.35 from Table 4.5. From Equation 1.1, the wind speed at a height of 2 meters is calculated to be 2.3 m s^{-1}. The instantaneous concentration χ_I at the fenceline, at any distance downwind, can be calculated from Equation 4.25 which with the effective height equal to 0, is:

$$\chi_I = 2\, Q_T/(15.75\ \sigma_x\ \sigma_y\ \sigma_z)$$

The dispersion parameters can be evaluated using the coefficients and exponents from Table 4.8 by:

$$\sigma_x = 0.045\ x^{0.91} \qquad \sigma_y = 0.045\ x^{0.91} \qquad \sigma_z = 0.12\ x^{0.67}$$

First, evaluating for the instantaneous concentration at a downwind distance of 400 m, $\sigma_x = \sigma_y = 10.5$ m, $\sigma_z = 6.64$ m.

$$\chi_I = 2\ 5000/(15.75\ \ 10.5\ \ 10.5\ \ 6.64) = 8.67 \times 10^{-1}\ \mathrm{g\ m^{-3}}$$

The highest 30-min average concentration when the center of the puff passes the point 400 m downwind halfway through the 30-min period can be determined by following the procedures in Section 4.17. Note that if we define the alongwind length of the puff as 8 σ_x, the length at the downwind distance of 400 m is only 84 m and passing at the speed of the wind of 2.3 m s^{-1} will pass by in only 36.5 seconds. Therefore the 30-minute concentration will be made up of zero concentrations for more than 29 minutes and non-zero concentrations for less than a minute.

Using Equation 4.26, N is:

$$N = \tau\, u/(2\ \sigma_x) = 1800\ \ 2.3/(2\ \ 10.5) = 197.1$$

Equation 4.27 is then used to determine F and since the entire puff passes by A = 1.:

$$F = (A - 0.5)/(0.3989\ N) = 0.5/78.6 = 6.36 \times 10^{-3}$$

The 30-min average concentration, χ_{30}, is then

$$\chi_{30} = \chi_I \times F = 8.67 \times 10^{-1}\ \ 6.36 \times 10^{-3}$$
$$= 5.52 \times 10^{-3}\ \mathrm{g\ m^{-3}} = 5520\ \mathrm{\mu g\ m^{-3}}$$

This is more than 4 times the LOC.

The distance to where the 30-min concentration falls to the LOC can be determined as follows:

Note that the instantaneous concentration at the downwind distance x can be determined as:

$$\chi_I = 10000/(15.75 \quad 0.045 \ x^{0.91} \ 0.045 \ x^{0.91} \ 0.12 \ x^{0.67})$$

$$= 2.61 \times 10^6/(x^{0.91} \ x^{0.91} \ x^{0.67})$$

The value for N is:

$$N = 1800 \ 2.3/(2 \ \sigma_x) = 2070/(0.045 \ x^{0.91}) = 46000/x^{0.91}$$

For the situations where the entire plume passes during the averaging time and therefore A = 1., when the above expression is substituted for N, F is:

$$F = 0.5/(0.3989 \ N) = 0.5 \ x^{0.91}/(0.3989 \ 46000)$$

$$= 2.725 \times 10^{-5} \quad x^{0.91}$$

The highest 30-min average concentration at the downwind distance x is then determined from:

$$\chi_{30} = \chi_I \times F = \frac{2.61 \times 10^6 \ 2.725 \times 10^{-5} \quad x^{0.91}}{x^{0.91} \quad x^{0.91} \quad x^{0.67}}$$

$$= 71.12/x^{1.58}$$

Substituting 1.25×10^{-3} g m^{-3} for χ_{30} and solving for x:

$$x = (71.12/1.25 \times 10^{-3})^{1/1.58} = (56896)^{0.633}$$

$$= 1023 \ m$$

Therefore, to our best estimate the highest 30-min average concentration would have exceeded the LOC at a distance of 1023 m from the accident site or 623 m beyond the nearest fenceline position. Note that depending upon the information available as to what the wind direction was last evening, we may be able to further pin down the area that was affected. If little information is known, then all we can say is that concentrations may have exceeded the LOC within a circle of 1023 m radius about the accident site.

Note: The program WKBK2 on the floppy diskette is not designed to make any calculations for line, area, or instantaneous sources.

REFERENCES

Benson, P. E., 1979: CALINE3 - A Versatile Dispersion Model for Predicting Air Pollutant Levels Near Highways and Arterial Streets. FHWA/CA/TL-79/23. Federal Highway Administration, Washington, D.C. (NTIS PB80-220 841).

Berkowicz, R., H. R. Olesen, and U. Torp, 1986: The Danish Gaussian air pollution model (OML): Description, test, and sensitivity analysis in view of regulatory applications. pp 453-481 in Air Pollution Modeling and Its Application V. (C. De Wispelaere, F. A. Schiermeier, and N. V. Gillani, Eds.) Plenum Press, New York.

Bierly, E. W., and E. W. Hewson, 1962: Some restrictive meteorological conditions to be considered in the design of stacks. *J. Appl. Meteorol.,* **1,** 3, 383-390.

Briggs, G. A., 1969: "Plume Rise," USAEC Critical Review Series, TID-25075, National Technical Information Service, Springfield, VA. 81 pp.

Briggs, G. A., 1971: Some recent analyses of plume rise observations. pp 1029-1032 in *Proceedings* of the Second International Clean Air Congress (H. M. Englund and W. T. Berry, Eds.) Academic Press, New York.

Briggs, G. A., 1972: Discussion on chimney plumes in neutral and stable surroundings. *Atmos. Environ.* **6,** 507-510.

Briggs, G. A., 1973: Diffusion Estimation for Small Emissions. Atmos. Turb. and Diff. Lab. Contribution File No. 79. Oak Ridge, TN. 59 pp.

Briggs, G. A., 1975: Chapter 3, Plume Rise Predictions. pp 59-111 in: Lectures on Air Pollution and Environmental Impact Analysis. (Duane A. Haugen, Ed.) Am. Meteorol. Soc. Boston, MA. 296 pp.

Briggs, G. A., 1984: Chapter 8, Plume Rise and Buoyancy Effects, pp 327-366 in "Atmospheric Science and Power Production" (Darryl Randerson, Ed.) DOE/TIC-27601. Technical Information Center. United States Department of Energy. [Available as DE84 005 177 from NTIS.]

Briggs, G. A., W. L. Eberhard, J. E. Gaynor, W. R. Moninger, and T. Uttal, 1986: Convective diffusion field measurements compared with laboratory and numerical experiments. pp. 340-343 in Fifth Joint Conference on Applications of Air Pollution Meteorology. November 18-21, 1986. Chapel Hill, NC. American Meteorological Society. Boston, MA.

Cramer, H. E., 1959: Engineering estimates of atmospheric dispersal capacity. *Am. Ind. Hyg. Assoc. J.,* **20,** 3, 183-189.

References.

Cramer, H. E., 1976: Improved techniques for modeling the dispersion of tall stack plumes. pp 731-780 in *Proceedings* of the 7th International Technical Meeting on Air Pollution Modeling and Its Application. September 7-10, 1976, Airlie, VA. NATO Committee on the Challenges to Modern Society No. 51. Brussels.

Davenport, A. G., 1963: The relationship of wind structure to wind loading. presented at Int. Conf. on The Wind Effects on Buildings and Structures, June 26-28, 1963. Natl. Physical Laboratory, Teddington, Middlesex, England.

Deardorff, J. W. and G. E. Willis, 1975: A parameterization of diffusion into the mixed layer. *J. Appl. Meteorol.,* **14,** 1451-1458.

Draxler, R. R., 1976: Determination of atmospheric diffusion parameters. *Atmos. Environ.,* **10,** 99-105.

Dyer, A. J., 1974: A review of flux-profile relationships. *Boundary-Layer Meteorol.,* **7,** 363-372.

Eberhard, W. L., W. R. Moninger, T. Uttal, S. W. Troxel, J. E. Gaynor, and G. A. Briggs, 1985: Field measurements in three dimensions of plume dispersion in the highly convective boundary layer. pp. 115-118. Seventh Symposium on Turbulence and Diffusion. November 12-15, 1985. Boulder, CO. American Meteorological Society. Boston, MA.

Environmental Protection Agency, 1986: Guideline on Air Quality Models (Revised). EPA-450/4-80-023R. Office of Air Quality Planning and Standards. Research Triangle Park, NC. (NTIS Accession Number PB86-245 248)

Environmental Protection Agency, 1987a: Industrial Source Complex (ISC) Dispersion Model User's Guide - Second Edition (Revised). Volumes 1 and 2. EPA-450/4-88-002a and b. Office of Air Quality Planning and Standards. Research Triangle Park, NC. (NTIS Accession Numbers: Vol. 1, PB88-171 475; Vol. 2, PB88-171 483.)

Environmental Protection Agency, 1987b: Supplement A to the Guideline on Air Quality Models (Revised). EPA-450/2-78-027R. Office of Air Quality Planning and Standards. Research Triangle Park, NC.

Farmer, S. F. G., 1983: Methods of estimating boundary layer depth. The Meteorological Office, Bracknell, U.K. 13 pp.

Gifford, F. A., 1959: Computation of pollution from several sources. *Int. J. Air Poll.,* **2,** 109-110.

Gifford, F. A., Jr., 1960a: Atmospheric dispersion calculations using the generalized Gaussian plume model. *Nuclear Safety,* **2** (2): 56-59, 67-68.

Gifford, F. A., 1960b: Peak to average concentration ratios according to a fluctuating plume dispersion model. *Int. J. Air Poll.,* **3,** 4, 253-260.

Gifford, F. A., 1962: The area within ground-level dosage isopleths. *Nuclear Safety,* **4,** 2, 91-92.

Gifford, F. A., 1976: Turbulent diffusion typing schemes: a review. *Nuclear Safety,* **17,** 1, 68-86.

Gryning, S. E., A. A. M. Holtslag, J. S. Irwin, and B. Sivertsen, 1987: Applied dispersion modeling based on meteorological scaling parameters. *Atmos. Environ.,* **21,** 1, 79-89.

Gryning, S. E., P. van Ulden, and S. E. Larson, 1983: Dispersion from a continuous ground-level source investigated by a K-model. *Quart. J. Roy. Meteorol. Soc.,* **109,** 355-364.

Hanna, Steven R., and Joseph C. Chang, 1990: Modification of the Hybrid Plume Dispersion Model (HPDM) for urban conditions and its evaluation using the Indianapolis data set. Report Number A089-1200.I, EPRI Project No. RP-02736-1. Prepared for The Electric Power Research Institute by Sigma Research Corporation, Westford MA.

Hanna, S. R., and J. C. Chang, 1992: Boundary-layer parameterizations for applied dispersion modeling over urban areas. *Boundary-Layer Meteorol.,* **58,** 229-259.

Hanna, Steven R. and Robert J. Paine, 1989: Hybrid Plume Dispersion Model (HPDM) development and evaluation. *J. Appl. Meteorol.,* **28,** 3, 206-224.

Hay, J. S., and F. Pasquill, 1959: Diffusion from a continuous source in relation to the spectrum and scale of turbulence. pp 345-365 in Atmospheric Diffusion and Air Pollution (F. N. Frenkiel and P. A. Sheppard, eds.) Advances in Geophysics, 6, Academic Press, New York, 471 pp.

Hewson, E. W., 1945: The meteorological control of atmospheric pollution by heavy industry. *Quart. J. Roy. Meteorol. Soc.,* **71,** 266-282.

Hewson, E. W., 1955: Stack heights required to minimize ground concentrations. *Trans. ASME,* **77,** 1163-1172.

Hewson, E. W., 1964: Industrial Air Pollution Meteorology. Meteorological Laboratories of the College of Engineering. The University of Michigan. Ann Arbor, MI. 191 pp.

Hewson, E. W., and G. C. Gill, 1944: Meteorological investigations in Columbia River Valley near Trail, B. C., pp. 23-228 in Report submitted to the Trail Smelter Arbitral Tribunal by R. S. Dean and R. E. Swain, Bur. of Mines Bull. 453, Govt. Print. Off., Washington, DC. 304 pp.

Hilsmeier, W. F., and F. A. Gifford, 1962: Graphs for estimating atmospheric diffusion. ORO-545, Atomic Energy Commission, Oak Ridge, TN., 10 pp.

References.

Hino, Mikio, 1968: Maximum ground-level concentration and sampling time. *Atmos. Environ.,* **2,** 149-165.

Holland, J. Z., 1953: A meteorological survey of the Oak Ridge area: Final report covering the period 1948-1952. Weather Bureau, pp. 554-559, U.S. Atomic Energy Comm. Report ORO-99, Washington, DC. 584 pp.

Holtslag, A. A. M., and A. P. van Ulden, 1983: A simple scheme for daytime estimates of the surface fluxes from routine weather data. *J. Climate Appl. Meteorol.,* **22,** 517-529.

Holzworth, G. C., 1972: Mixing Heights, Wind Speeds, and Potential for Urban Air Pollution throughout the Contiguous United States. AP-101, U.S. Environmental Protection Agency, Research Triangle Park, NC. 118 pp.

Huang, C. H., 1979: A theory of dispersion in turbulent shear flow. *Atmos. Environ.,* **13,** 453-461.

Irwin, J. S., 1983: Estimating plume dispersion -- a comparison of several sigma schemes. *J. Climate Appl. Meteorol.,* **22,** 92-114.

Lamb, R. G., 1982: Diffusion in the convective boundary layer. Chapter 5. pp 159-229 in Atmospheric Turbulence and Air Pollution Modeling. F. T. M. Nieuwstadt and H. van Dop, Eds., Reidel, Dordrecht, Holland.

McElroy, J. L., and F. Pooler, 1968: St. Louis Dispersion Study. U.S. Public Health Service, National Air Pollution Control Administration Report AP-53.

Moninger, W. R., W. L. Eberhard, G. A. Briggs, R. A. Kropfli, and J. C. Kaimal, 1983: Simultaneous radar and lidar observations of plumes from continuous point sources. pp. 246-250 in 21st Radar Meteorology Conference, American Meteorological Society, Boston, MA.

Nieuwstadt, F. T. M., and A. P. van Ulden, 1978: A numerical study on the vertical dispersion of passive contaminants from a continuous source in the atmospheric surface layer. *Atmos. Environ.,* **12,** 2119-2124.

Nonhebel, G., 1960: Recommendations on heights for new industrial chimneys. *J. Inst. Fuel,* **33,** 479-513.

Pasquill, F., 1961: The estimation of the dispersion of windborne material. *Meteorol. Mag.,* **90** (1063): 33-49.

Pasquill, F., 1976: Atmospheric Dispersion Parameters in Gaussian Plume Modeling: Part II. Possible Requirements for Change in the Turner Workbook Values. EPA-600/4-76-030b. U.S. Environmental Protection Agency. Research Triangle Park, NC, 44pp.

Petersen, W. B., 1980: User's Guide for HIWAY-2. EPA-600/8-80-018. U.S. Environmental Protection Agency, Research Triangle Park, NC. (NTIS PB80-227 556).

Petersen, W. B., 1982: Estimating Concentrations Downwind from an Instantaneous Puff Release. EPA-600/3-82-078. U.S. Environmental Protection Agency, Research Triangle Park, NC. 73 pp.

Petersen, W. B., and E. D. Rumsey, 1987: User's Guide for PAL 2.0 -- A Gaussian-Plume Algorithm for Point, Area, and Line Sources. EPA/600/8-87/009. U.S. Environmental Protection Agency, Research Triangle Park, NC. (NTIS PB87-168 787).

Pierce, Thomas E., D. Bruce Turner, Joseph A. Catalano, and Frank V. Hale III, 1982: PTPLU - A Single Source Gaussian Dispersion Algorithm - User's Guide," EPA-600/8-82-014, U. S. Environmental Protection Agency, Research Triangle Park, NC.

Pooler, F., 1965: Potential dispersion of plumes from large power plants. PHS Publ. No. 999-AP-16. 13 pp.

Singer, I. A., 1961: The relation between peak and mean concentrations. *J. Air Poll. Control Assoc.,* **11,** 336-341.

Singer, I. A., K. Imai, and R. G. Del Campos, 1963: Peak to mean pollutant concentration ratios for various terrain and vegetation cover. *J. Air Poll. Control Assoc.,* **13,** 40-42.

Singer, I. A., and M. E. Smith, 1953: Relation of gustiness to other meteorological parameters. *J. Meteorol.,* **10,** 121-126.

Slade, D. H., 1965: Dispersion estimates from pollutant releases of a few seconds to 8 hours in duration. Unpublished Weather Bureau Report. Aug. 1965.

Smith, F. B., 1973: A scheme for estimating the vertical dispersion of a plume from a source near ground level. Chap. 17 in NATO CCMS Air Pollution No. 14, Paris, France.

Smith, F. B., 1983: Program to determine boundary layer parameters. Manuscript 14, Meteorological Office, Bracknell, U. K.

Smith, M. E., 1963: The use and misuse of the atmosphere. Brookhaven Lecture Series, No. 24, February 13, 1963, BNL 784 (T-298). Brookhaven National Laboratory. 15 pp.

Stewart, N. G., H.J. Gale, and R. N. Crooks, 1958: The atmospheric diffusion of gases discharged from the chimney of the Harwell Reactor BEPO. *Int. J. Air Poll.,* **1,** 87-102.

Sutton, O. G., 1932: A theory of eddy diffusion in the atmosphere. *Proc. Roy. Soc.,* **A,** 135, 143-165.

Taylor, G. I., 1915: Eddy motion in the atmosphere. *Phil. Trans. Roy. Soc.,* **A,** 215, 1-26.

Turner, D. Bruce, 1967: Workbook of Atmospheric Dispersion Estimates, PHS Publ. No. 999 AP-26, Cincinnati, Ohio, 84 pp.

References.

Turner, D. B., 1985: Proposed pragmatic methods for estimating plume rise and plume penetration through atmospheric layers, Preliminary Communication. *Atmos. Environ.* **19** (7) 1215-1218.

Turner, D. Bruce, 1986: Addendum to "TUPOS - Incorporation of a hesitant plume algorithm," EPA-600/8-86/027, U. S. Environmental Protection Agency, Research Triangle Park, NC.

Turner, D. Bruce, Lucille W. Bender, James O. Paumier and Phillip F. Boone, 1991: Evaluation of TUPOS air quality model using data from the EPRI Kincaid field study. *Atmos. Environ.,* **25A,** 10, 2187-2201.

Turner, D. Bruce, and Joan H. Novak, 1978: User's Guide for RAM, Volume I, Algorithm Description and Use, (60 pp), Volume II, Data Preparation and Listings, (222 pp), EPA-600/8-78-016 a and b. U. S. Environmental Protection Agency, Research Triangle Park, NC.

Weil, J. C., 1988: Plume Rise, pp 119-166 in "Lectures on Air Pollution Modeling" (A. Venkatram and J. C. Wyngaard, Eds.) Amer. Meteorol. Soc. Boston.

Weil, J. C., and R. P. Brower, 1983: Estimating Convective Boundary Layer Parameters for Diffusion Applications. PPSP-MP-48. Environmental Center, Martin Marietta Corp. for Maryland Department of Natural Resources. Baltimore, MD. 45 pp.

Weil, J. C. and R. P. Brower, 1984: An updated Gaussian plume model for tall stacks. *J. Air Poll. Control Assoc.,* **34,** (8) 818-827.

Weil, J. C., and L. A. Corio, 1988: A modification of the PPSP dispersion model for highly buoyant plumes. Report PPRP-MP-60. Maryland Power Plant Research Program, Maryland Department of Natural Resources, Annapolis, MD.

Willis, G. E. and J. W. Deardorff, 1978: A laboratory study of dispersion from an elevated source within a modeled convective boundary layer. *Atmos. Environ.,* **12,** 1305-1311.

Wyngaard, J. C., 1988: Structure of the PBL. pp 9-62 in "Lectures on Air Pollution Modeling." Amer. Meteorol. Soc. Boston, MA.

APPENDIX

ABBREVIATIONS

cal	calorie
g	gram
°K	degrees Kelvin
m	meter
mb	millibar
sec	second

SYMBOLS

c_p	specific heat at constant pressure
d	inside stack top diameter
D_T	Total Dosage
e	2.7183
$f(\theta,S,N)$	frequency of wind direction for a given stability and wind speed class
h	physical stack height
H	effective height of emission
H_s	surface heat flux
k	von Karman's constant, approximately equal to 0.4
L	Monin-Obukhov length
p	two uses: atmospheric pressure
	wind speed profile exponent
q	emission rate per length of a line source
Q	emission rate for a point source
R	net rate of sensible heating of an air column by solar radiation
t_k	a shorttime period
t_s	a time period
T	air temperature
T_s	stack gas exit temperature
u	wind speed
v_s	stack gas exit velocity
x	downwind distance
y	crosswind distance
z	height above ground level
z_o	roughness length
$\partial\theta/\partial z$	the rate of change of potential temperature with height
ΔH	the rise of the plume centerline above the stack top
θ	potential temperature
σ_a	the standard deviation of wind azimuth (direction) as determined from a wind vane or bi-directional vane
σ_e	the standard deviation of wind elevation angle

Appendix

σ_x	the standard deviation in the downwind direction of a puff concentration distribution
σ_y	the standard deviation in the crosswind direction of the plume concentration distribution
σ_z	the standard deviation in the vertical of the plume concentration distribution
χ	concentration
χ/Q	relative concentration
$\chi u/Q$	relative concentration normalized for wind speed

CONSTANTS

e	=	2.7183	$1/e$	=	0.3679
π	=	3.1416	$1/\pi$	=	0.3183
2π	=	6.2832	$1/(2\pi)$	=	0.1592
$(2\pi)^{0.5}$	=	2.5066	$1/[(2\pi)^{0.5}]$	=	0.3989
			$2/[(2\pi)^{0.5}]$	=	0.7979
$(2\pi)^{3/2}$	=	15.75			

TEMPERATURE CONVERSION EQUATIONS

$$T(°C) \quad = \quad 5/9 \; [T(°F) - 32]$$

$$T(°K) \quad = \quad T(°C) \quad + \quad 273.16$$

$$T(°F) \quad = \quad [9/5[T(°C)] + \quad 32$$

CHARACTERISTICS OF THE GAUSSIAN DISTRIBUTION

The Gaussian or normal distribution can be depicted by the bell-shaped curve shown in Figure 2.1. The equation for the ordinate value of this curve is:

$$y \quad = \quad \frac{1}{(2\pi)^{0.5} \, \sigma} \left[-\frac{1}{2} \left(\frac{x - \mu}{\sigma} \right)^2 \right]$$

Figure A.1 gives the ordinate value at any distance from the center of the distribution (which occurs at μ). This information is also given in Table 2.1 Figure A.2 gives the area under the Gaussian curve from $-\infty$ to a particular value of p where $p \; = \; (x - \mu)/\sigma$.

This area is found from:

$$\text{Area } (-\infty \text{ to } p) \quad = \quad \int_{-\infty}^{p} \frac{1}{(2\pi)^{0.5}} \; \exp{(-0.5 \, p^2)} \quad dp$$

A-2

Figure A.3 gives the area under the Gaussian curve from –p to +p. This can be found from:

$$\text{Area } (-p \text{ to } p) = \int_{-p}^{+p} \frac{1}{(2\pi)^{0.5}} \exp(-0.5\, p^2) \; dp$$

CONVERSION TABLES

The following tables can be used to find the conversion factor to convert from given units to desired units. For the row with the given units stated in the left column of the table find the multiplier in this row beneath the column heading with the desired units. Multiplying a value with the given units by this multiplier will convert it to a value with the desired units.

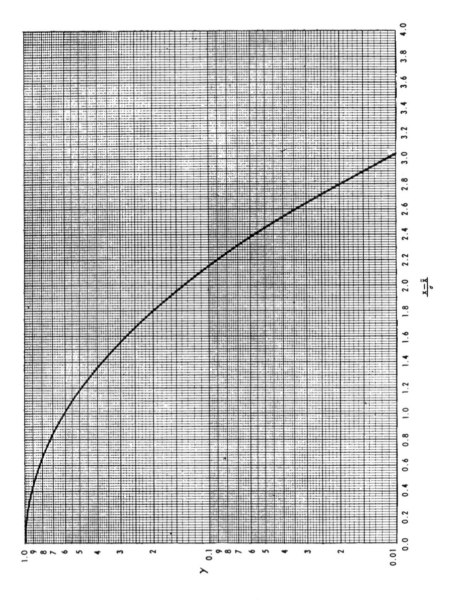

Figure A.1 Ordinate values of the Gaussian distribution.

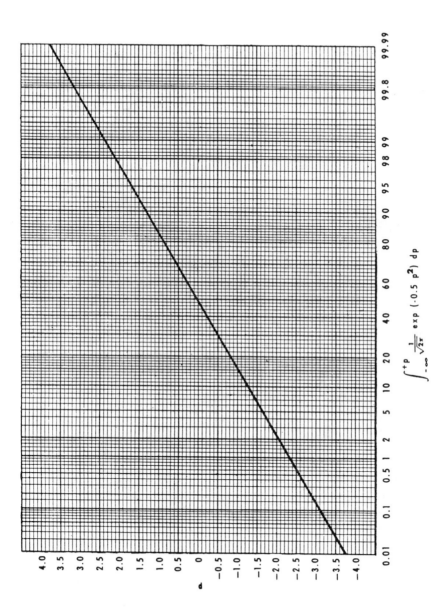

Figure A.2   Area under the Gaussian distribution curve from -∞ to p.

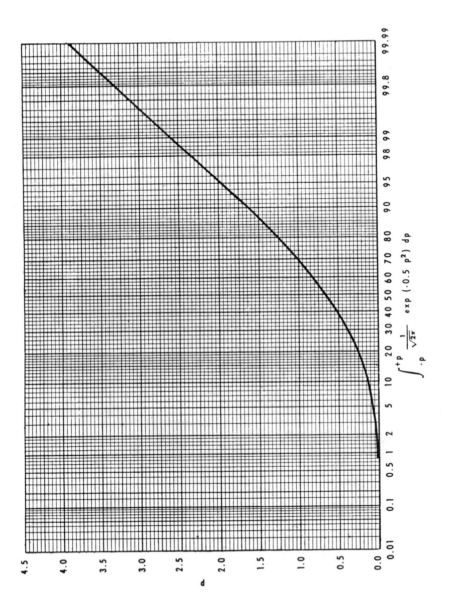

Figure A.3 Area under the Gaussian distribution curve between -p and +p.

CONVERSION FACTORS - VELOCITY

DESIRED UNITS:

GIVEN UNITS	METERS PER SEC	KM PER HR	KNOTS	MI(STAT) PER HR	FEET PER SEC	FEET PER MIN	MI(STAT) PER DAY	CM PER SEC	METERS PER MIN
METERS PER SEC	1.0000E+00	3.6000E+00	1.9440E+00	2.2371E+00	3.2808E+00	1.9685E+02	5.3686E+01	1.0000E+02	5.9988E+01
KM PER HR	2.7778E-01	1.0000E+00	5.4001E-01	6.2143E-01	9.1135E-01	5.4681E+01	1.4913E+01	2.7778E+01	1.6663E+01
KNOTS	5.1440E-01	1.8518E+00	1.0000E+00	1.1508E+00	1.6877E+00	1.0126E+02	2.7616E+01	5.1440E+01	3.0858E+01
MI(STAT) PER HR	4.4700E-01	1.6092E+00	8.6897E-01	1.0000E+00	1.4665E+00	8.7992E+01	2.3997E+01	4.4700E+01	2.6815E+01
FEET PER SEC	3.0480E-01	1.0973E+00	5.9253E-01	6.8188E-01	1.0000E+00	6.0000E+01	1.6363E+01	3.0480E+01	1.8284E+01
FEET PER MIN	5.0800E-03	1.8288E-02	9.8756E-03	1.1365E-02	1.6667E-02	1.0000E+00	2.7272E-01	5.0800E-01	3.0474E-01
MI(STAT) PER DAY	1.8627E-02	6.7057E-02	3.6211E-02	4.1671E-02	6.1112E-02	3.6667E+00	1.0000E+00	1.8627E+00	1.1174E+00
CM PER SEC	1.0000E-02	3.6000E-02	1.9440E-02	2.2371E-02	3.2808E-02	1.9685E+00	5.3686E-01	1.0000E+00	5.9988E-01
METERS PER MIN	1.6670E-02	6.0012E-02	3.2407E-02	3.7293E-02	5.4692E-02	3.2815E+00	8.9494E-01	1.6670E+00	1.0000E+00

CONVERSION FACTORS - EMISSION RATES

DESIRED UNITS:

GIVEN UNITS	GRAMS PER SEC	GRAMS PER MIN	KG PER HOUR	KG PER DAY	LBS PER MIN	LBS PER HOUR	LBS PER DAY	TONS PER HOUR	TONS PER DAY
GRAMS PER SEC	1.0000E+00	5.9999E+01	3.6000E+00	8.6401E+01	1.3228E-01	7.9365E+00	1.9048E+02	3.9683E-03	9.5238E-02
GRAMS PER MIN	1.6667E-02	1.0000E+00	6.0001E-02	1.4400E+00	2.2047E-03	1.3228E-01	3.1747E+00	6.6139E-05	1.5873E-03
KG PER HOUR	2.7778E-01	1.6666E+01	1.0000E+00	2.4000E+01	3.6744E-02	2.2046E+00	5.2911E+01	1.1023E-03	2.6455E-02
KG PER DAY	1.1574E-02	6.9443E-01	4.1666E-02	1.0000E+00	1.5310E-03	9.1857E-02	2.2046E+00	4.5929E-05	1.1023E-03
LBS PER MIN	7.5599E+00	4.5358E+02	2.7215E+01	6.5318E+02	1.0000E+00	5.9999E+01	1.4400E+03	3.0000E-02	7.1999E-01
LBS PER HOUR	1.2600E-01	7.5598E+00	4.5360E-01	1.0886E+01	1.6667E-02	1.0000E+00	2.4000E+01	5.0000E-04	1.2000E-02
LBS PER DAY	5.2499E-03	3.1499E-01	1.8899E-02	4.5359E-01	6.9444E-04	4.1666E-02	1.0000E+00	2.0833E-05	4.9999E-04
TONS PER HOUR	2.5200E+02	1.5120E+04	9.0719E+02	2.1773E+04	3.3334E+01	2.0000E+03	4.8001E+04	1.0000E+00	2.4000E+01
TONS PER DAY	1.0500E+01	6.2999E+02	3.7800E-01	9.0721E+02	1.3889E+00	8.3333E+01	2.0000E+03	4.1667E-02	1.0000E+00

CONVERSION FACTORS - LENGTH

DESIRED UNITS:

GIVEN UNITS	METER	CM	MICROMETER	KILOMETER	INCH	FOOT	YARD	MILE(STAT)	MILE(NAUT)
METER	1.0000E+00	1.0000E+02	1.0000E+06	1.0000E-03	3.9370E+01	3.2808E+00	1.0936E+00	6.2139E-04	5.3961E-04
CM	1.0000E-02	1.0000E+00	1.0000E+04	1.0000E-05	3.9370E-01	3.2808E-02	1.0936E-02	6.2139E-06	5.3961E-06
MICROMETER	1.0000E-06	1.0000E-04	1.0000E+00	1.0000E-09	3.9370E-05	3.2808E-06	1.0936E-06	6.2139E-10	5.3961E-10
KILOMETER	1.0000E+03	1.0000E+05	1.0000E+09	1.0000E+00	3.9370E+04	3.2808E+03	1.0936E+03	6.2139E-01	5.3961E-01
INCH	2.5400E-02	2.5400E+00	2.5400E+04	2.5400E-05	1.0000E+00	8.3333E-02	2.7778E-02	1.5783E-05	1.3706E-05
FOOT	3.0480E-01	3.0480E+01	3.0480E+05	3.0480E-04	1.2000E+01	1.0000E+00	3.3333E-01	1.8940E-04	1.6447E-04
YARD	9.1440E-01	9.1440E+01	9.1440E+05	9.1440E-04	3.6000E+01	3.0000E+00	1.0000E+00	5.6820E-04	4.9342E-04
MILE(STAT)	1.6093E+03	1.6093E+05	1.6093E+09	1.6093E+00	6.3358E+04	5.2799E+03	1.7600E+03	1.0000E+00	8.6839E-01
MILE(NAUT)	1.8532E+03	1.8532E+05	1.8532E+09	1.8532E+00	7.2961E+04	6.0801E+03	2.0267E+03	1.1516E+00	1.0000E+00

CONVERSION FACTORS - AREA

DESIRED UNITS:

GIVEN UNITS	SQUARE METER	SQUARE KM	SQUARE CM	SQUARE INCH	SQUARE FOOT	SQUARE YARD	ACRE	SQUARE STAT MI	SQUARE NAUT MI
SQUARE METER	1.0000E+00	1.0000E-06	1.0000E+04	1.5500E+03	1.0764E+01	1.1960E+00	2.4710E-04	3.8610E-07	2.9117E-07
SQUARE KM	1.0000E+06	1.0000E+00	1.0000E+10	1.5500E+09	1.0764E+07	1.1960E+06	2.4710E+02	3.8610E-01	2.9117E-01
SQUARE CM	1.0000E-04	1.0000E-10	1.0000E+00	1.5500E-01	1.0764E-03	1.1960E-04	2.4710E-08	3.8610E-11	2.9117E-11
SQUARE INCH	6.4516E-04	6.4516E-10	6.4516E+00	1.0000E+00	6.9444E-03	7.7160E-04	1.5942E-07	2.4910E-10	1.8785E-10
SQUARE FOOT	9.2903E-02	9.2903E-08	9.2903E+02	1.4400E+02	1.0000E+00	1.1111E-01	2.2957E-05	3.5870E-08	2.7051E-08
SQUARE YARD	8.3613E-01	8.3613E-07	8.3613E+03	1.2960E+03	9.0000E+00	1.0000E+00	2.0661E-04	3.2283E-07	2.4346E-07
ACRE	4.0469E+03	4.0469E-03	4.0469E+07	6.2727E+06	4.3560E+04	4.8400E+03	1.0000E+00	1.5625E-03	1.1783E-03
SQUARE STAT MI	2.5900E+06	2.5900E+00	2.5900E+10	4.0145E+09	2.7879E+07	3.0976E+06	6.4000E+02	1.0000E+00	7.5413E-01
SQUARE NAUT MI	3.4344E+06	3.4344E+00	3.4344E+10	5.3233E+09	3.6968E+07	4.1075E+06	8.4865E+02	1.3260E+00	1.0000E+00

CONVERSION FACTORS - VOLUME

DESIRED UNITS:

GIVEN UNITS	CUBIC METER	CUBIC INCH	CUBIC FOOT	CUBIC YARD	U. S. QT	FLUID OZ	U. S. GALLON	U. S. BUSHELS	LITERS
CUBIC METER	1.0000E+00	6.1024E+04	3.5314E+01	1.3079E+00	1.0567E+03	3.3813E+04	2.6417E+02	2.8391E+01	1.0000E+03
CUBIC INCH	1.6387E-05	1.0000E+00	5.7870E-04	2.1433E-05	1.7316E-02	5.5410E-01	4.3290E-03	4.6524E-04	1.6387E-02
CUBIC FOOT	2.8317E-02	1.7280E+03	1.0000E+00	3.7037E-02	2.9922E+01	9.5750E+02	7.4806E+00	8.0393E-01	2.8317E+01
CUBIC YARD	7.6456E-01	4.6656E+04	2.7000E+01	1.0000E+00	8.0790E+02	2.5852E+04	2.0198E+02	2.1706E+01	7.6456E+02
U. S. QT	9.4635E-04	5.7750E+01	3.3420E-02	1.2378E-03	1.0000E+00	3.1999E+01	2.5000E-01	2.6867E-02	9.4635E-01
FLUID OZ	2.9574E-05	1.8047E+00	1.0444E-03	3.8681E-05	3.1251E-02	1.0000E+00	7.8126E-03	8.3962E-04	2.9574E-02
U. S. GALLON	3.7854E-03	2.3100E+02	1.3368E-01	4.9511E-03	4.0000E+00	1.2800E+02	1.0000E+00	1.0747E-01	3.7854E+00
U. S. BUSHELS	3.5223E-02	2.1494E+03	1.2439E+00	4.6070E-02	3.7220E+01	1.1910E+03	9.3050E+00	1.0000E+00	3.5223E+01
LITERS	1.0000E-03	6.1024E+01	3.5314E-02	1.3079E-03	1.0567E+00	3.3813E+01	2.6417E-01	2.8391E-02	1.0000E+00

CONVERSION FACTORS - MASS

DESIRED UNITS:

GIVEN UNITS	KILOGRAMS	LBS (TROY)	LBS (AVOIR)	GRAINS	OZ (TROY)	OZ (AVOIR)	TONS SHORT	TONS LONG	TONS METRIC
KILOGRAMS	1.0000E+00	2.6795E+00	2.2046E+00	1.5432E+04	3.2154E+01	3.5275E+01	1.1023E-03	9.8425E-04	1.0000E-03
LBS (TROY)	3.7320E-01	1.0000E+00	8.2277E-01	5.7593E+03	1.2000E+01	1.3164E+01	4.1138E-04	3.6732E-04	3.7320E-04
LBS (AVOIR)	4.5359E-01	1.2154E+00	1.0000E+00	7.0000E+03	1.4585E+01	1.6000E+01	5.0000E-04	4.4645E-04	4.5359E-04
GRAINS	6.4799E-05	1.7363E-04	1.4286E-04	1.0000E+00	2.0836E-03	2.2858E-03	7.1429E-08	6.3779E-08	6.4799E-08
OZ (TROY)	3.1100E-02	8.3333E-02	6.8564E-02	4.7995E+02	1.0000E+00	1.0970E+00	3.4282E-05	3.0610E-05	3.1100E-05
OZ (AVOIR)	2.8349E-02	7.5962E-02	6.2499E-02	4.3749E+02	9.1154E-01	1.0000E+00	3.1250E-05	2.7903E-05	2.8349E-05
TONS SHORT	9.0718E+02	2.4308E+03	2.0000E+03	1.4000E+07	2.9170E+04	3.2000E+04	1.0000E+00	8.9289E-01	9.0718E-01
TONS LONG	1.0160E+03	2.7224E+03	2.2399E+03	1.5679E+07	3.2669E+04	3.5839E+04	1.1200E+00	1.0000E+00	1.0160E+00
TONS METRIC	1.0000E+03	2.6795E+03	2.2046E+03	1.5432E+07	3.2154E+04	3.5275E+04	1.1023E+00	9.8425E-01	1.0000E+00

CONVERSION FACTORS - FLOW

DESIRED UNITS:

GIVEN UNITS	CUBIC M PER SEC	CUBIC FT PER SEC	CUBIC FT PER MIN	CUBIC FT PER HR	LITER PER SEC	LITER PER MIN	LITER PER HR	CUBIC M PER HOUR	CUBIC CM PER SEC
CUBIC M PER SEC	1.0000E+00	3.5314E+01	2.1189E+03	1.2713E+05	1.0000E+03	5.9999E+04	3.6000E+06	3.6000E+03	1.0000E+06
CUBIC FT PER SEC	2.8317E-02	1.0000E+00	6.0000E+01	3.6000E+03	2.8317E+01	1.6990E+03	1.0194E+05	1.0194E+02	2.8317E+04
CUBIC FT PER MIN	4.7195E-04	1.6667E-02	1.0000E+00	6.0000E+01	4.7195E-01	2.8316E+01	1.6990E+03	1.6990E+00	4.7195E+02
CUBIC FT PER HR	7.8658E-06	2.7778E-04	1.6667E-02	1.0000E+00	7.8658E-03	4.7194E-01	2.8317E+01	2.8317E-02	7.8658E+00
LITER PER SEC	1.0000E-03	3.5314E-02	2.1189E+00	1.2713E+02	1.0000E+00	5.9999E+01	3.6000E+03	3.6000E+00	1.0000E+03
LITER PER MIN	1.6667E-05	5.8859E-04	3.5315E-02	2.1189E+00	1.6667E-02	1.0000E+00	6.0001E+01	6.0001E-02	1.6667E+01
LITER PER HR	2.7778E-07	9.8096E-06	5.8858E-04	3.5315E-02	2.7778E-04	1.6666E-02	1.0000E+00	9.9999E-04	2.7778E-01
CUBIC M PER HOUR	2.7778E-04	9.8097E-03	5.8858E-01	3.5315E+01	2.7778E-01	1.6666E+01	1.0000E+03	1.0000E+00	2.7778E+02
CUBIC CM PER SEC	1.0000E-06	3.5314E-05	2.1189E-03	1.2713E-01	1.0000E-03	5.9999E-02	3.6000E+00	3.6000E-03	1.0000E+00

CONVERSION FACTORS - DENSITY

DESIRED UNITS:

GIVEN UNITS	GRAMS PER CUBIC M	MGRAM PER CUBIC M	MICROGM PER CUB M	MICROGM PER LITER	GRAIN PER CUBIC FT	OZ PER CUBIC FT	LB PER CUBIC FT	GRAM PER CUBIC FT	LB PER CUBIC M
GRAMS PER CUBIC M	1.0000E+00	1.0000E+03	1.0000E+06	1.0000E+03	4.3701E-01	9.9890E-04	6.2430E-05	2.8317E-02	2.2046E-03
MGRAM PER CUBIC M	1.0000E-03	1.0000E+00	1.0000E+03	1.0000E+00	4.3701E-04	9.9890E-07	6.2430E-08	2.8317E-05	2.2046E-06
MICROGM PER CUB M	1.0000E-06	1.0000E-03	1.0000E+00	1.0000E-03	4.3701E-07	9.9890E-10	6.2430E-11	2.8317E-08	2.2046E-09
MICROGM PER LITER	1.0000E-03	1.0000E+00	1.0000E+03	1.0000E+00	4.3701E-04	9.9890E-07	6.2430E-08	2.8317E-05	2.2046E-06
GRAIN PER CUBIC FT	2.2883E+00	2.2883E+03	2.2883E+06	2.2883E+03	1.0000E+00	2.2858E-03	1.4286E-04	6.4799E-02	5.0449E-03
OZ PER CUBIC FT	1.0011E+03	1.0011E+06	1.0011E+09	1.0011E+06	4.3749E+02	1.0000E+00	6.2498E-02	2.8349E+01	2.2071E+00
LB PER CUBIC FT	1.6018E+04	1.6018E+07	1.6018E+10	1.6018E+07	7.0000E+03	1.6000E+01	1.0000E+00	4.5359E+02	3.5314E+01
GRAM PER CUBIC FT	3.5314E+01	3.5314E+04	3.5314E+07	3.5314E+04	1.5432E+01	3.5275E-02	2.2046E-03	1.0000E+00	7.7854E-02
LB PER CUBIC M	4.5359E+02	4.5359E+05	4.5359E+08	4.5359E+05	1.9822E+02	4.5309E-01	2.8318E-02	1.2844E+01	1.0000E+00

CONVERSION FACTORS - DEPOSITION

DESIRED UNITS:

GIVEN UNITS	GRAMS PER SQ M-MO	MG PER SQ CM-MO	KG PER SQ KM-MO	TON PER SQ MI-MO	OZ PER SQ FT-MO	LBS PER SQ FT-MO	LBS PER ACRE-MO	GRAMS PER SQ FT-MO	MG PER SQ IN-MO
GRAMS PER SQ M-MO	1.0000E+00	1.0000E-01	1.0000E+03	2.8549E+00	3.2771E-03	2.0482E-04	8.9222E+00	9.2902E-02	6.4516E-01
MG PER SQ CM-MO	1.0000E+01	1.0000E+00	1.0000E+04	2.8549E+01	3.2771E-02	2.0482E-03	8.9222E+01	9.2902E-01	6.4516E+00
KG PER SQ KM-MO	1.0000E-03	1.0000E-04	1.0000E+00	2.8549E-03	3.2771E-06	2.0482E-07	8.9222E-03	9.2902E-05	6.4516E-04
TON PER SQ MI-MO	3.5028E-01	3.5028E-02	3.5028E+02	1.0000E+00	1.1479E-03	7.1743E-05	3.1253E+00	3.2542E-02	2.2599E-01
OZ PER SQ FT-MO	3.0515E+02	3.0515E+01	3.0515E+05	8.7116E+02	1.0000E+00	6.2500E-02	2.7226E+03	2.8349E+01	1.9687E+02
LBS PER SQ FT-MO	4.8824E+03	4.8824E+02	4.8824E+06	1.3939E+04	1.6000E+01	1.0000E+00	4.3562E+04	4.5359E+02	3.1499E+03
LBS PER ACRE-MO	1.1208E-01	1.1208E-02	1.1208E+02	3.1997E-01	3.6729E-04	2.2956E-05	1.0000E+00	1.0412E-02	7.2310E-02
GRAMS PER SQ FT-MO	1.0764E+01	1.0764E+00	1.0764E+04	3.0730E+01	3.5274E-02	2.2047E-03	9.6039E+01	1.0000E+00	6.9445E+00
MG PER SQ IN-MO	1.5500E+00	1.5500E-01	1.5500E+03	4.4250E+00	5.0795E-03	3.1747E-04	1.3829E+01	1.4400E-01	1.0000E+00

CONVERSION FACTORS - PRESSURE

DESIRED UNITS:

GIVEN UNITS	MILLIBAR	PA PASCALS	POUNDS PER SQ IN	BAR	ATMOSPHERE	MM MERCURY (0 C)	IN MERCURY (0 C)	KG PER SQ CM	INCH WATER (15 C)
MILLIBAR	1.0000E+00	1.0000E+02	1.4504E-02	1.0000E-03	9.8687E-04	7.5008E-01	2.9530E-02	1.0197E-03	4.0177E-01
PA PASCALS	1.0000E-02	1.0000E+00	1.4504E-04	1.0000E-05	9.8687E-06	7.5008E-03	2.9530E-04	1.0197E-05	4.0177E-03
POUNDS PER SQ IN	6.8947E+01	6.8947E+03	1.0000E+00	6.8947E-02	6.8042E-02	5.1715E+01	2.0360E+00	7.0307E-02	2.7701E+01
BAR	1.0000E+03	1.0000E+05	1.4504E+01	1.0000E+00	9.8687E-01	7.5008E+02	2.9530E+01	1.0197E+00	4.0177E+02
ATMOSPHERE	1.0133E+03	1.0133E+05	1.4697E+01	1.0133E+00	1.0000E+00	7.6005E+02	2.9923E+01	1.0333E+00	4.0711E+02
MM MERCURY (0 C)	1.3332E+00	1.3332E+02	1.9337E-02	1.3332E-03	1.3157E-03	1.0000E+00	3.9369E-02	1.3595E-03	5.3564E-01
IN MERCURY (0 C)	3.3864E+01	3.3864E+03	4.9116E-01	3.3864E-02	3.3420E-02	2.5401E+01	1.0000E+00	3.4532E-02	1.3605E+01
KG PER SQ CM	9.8066E+02	9.8066E+04	1.4223E+01	9.8066E-01	9.6779E-01	7.3557E+02	2.8959E+01	1.0000E+00	3.9400E+02
INCH WATER (15 C)	2.4890E+00	2.4890E+02	3.6100E-02	2.4890E-03	2.4563E-03	1.8669E+00	7.3500E-02	2.5381E-03	1.0000E+00

CONVERSION FACTORS - TIME

DESIRED UNITS:

GIVEN UNITS	SECONDS	MINUTES	HOUR	DAY	WEEK	MONTH (30)	MONTH (31)	YEAR (365)	YEAR (366)
SECONDS	1.0000E+00	1.6667E-02	2.7778E-04	1.1574E-05	1.6534E-06	3.8580E-07	3.7336E-07	3.1710E-08	3.1623E-08
MINUTES	6.0000E+01	1.0000E+00	1.6667E-02	6.9444E-04	9.9206E-05	2.3148E-05	2.2401E-05	1.9026E-06	1.8974E-06
HOUR	3.6000E+03	6.0000E+01	1.0000E+00	4.1667E-02	5.9524E-03	1.3889E-03	1.3441E-03	1.1416E-04	1.1384E-04
DAY	8.6400E+04	1.4400E+03	2.4000E+01	1.0000E+00	1.4286E-01	3.3333E-02	3.2258E-02	2.7397E-03	2.7322E-03
WEEK	6.0480E+05	1.0080E+04	1.6800E+02	7.0000E+00	1.0000E+00	2.3333E-01	2.2581E-01	1.9178E-02	1.9126E-02
MONTH (30)	2.5920E+06	4.3200E+04	7.2000E+02	3.0000E+01	4.2857E+00	1.0000E+00	9.6774E-01	8.2192E-02	8.1967E-02
MONTH (31)	2.6784E+06	4.4640E+04	7.4400E+02	3.1000E+01	4.4286E+00	1.0333E+00	1.0000E+00	8.4932E-02	8.4699E-02
YEAR (365)	3.1536E+07	5.2560E+05	8.7600E+03	3.6500E+02	5.2143E+01	1.2167E+01	1.1774E+01	1.0000E+00	9.9727E-01
YEAR (366)	3.1622E+07	5.2704E+05	8.7840E+03	3.6600E+02	5.2286E+01	1.2200E+01	1.1806E+01	1.0027E+00	1.0000E+00

CONVERSION FACTORS - POWER

DESIRED UNITS:

GIVEN UNITS	WATT (INT)	KILOWATT (INT)	MEGAWATT (INT)	CAL (INT) PER SEC	BTU PER SEC	BTU PER MIN	JOULES ABS PER SEC	WATTS (ABS)	HORSEPWR (ELECT)
WATT (INT)	1.0000E+00	1.0000E-03	1.0000E-06	2.3880E-01	9.4845E-04	5.6907E-02	1.0002E+00	1.0002E+00	1.3407E-03
KILOWATT (INT)	1.0000E+03	1.0000E+00	1.0000E-03	2.3880E+02	9.4845E-01	5.6907E+01	1.0002E+03	1.0002E+03	1.3407E+00
MEGAWATT (INT)	1.0000E+06	1.0000E+03	1.0000E+00	2.3880E+05	9.4845E+02	5.6907E+04	1.0002E+06	1.0002E+06	1.3407E+03
CAL (INT) PER SEC	4.1876E+00	4.1876E-03	4.1876E-06	1.0000E+00	3.9717E-03	2.3830E-01	4.1884E+00	4.1884E+00	5.6145E-03
BTU PER SEC	1.0543E+03	1.0544E+00	1.0543E-03	2.5178E+02	1.0000E+00	6.0000E+01	1.0546E+03	1.0546E+03	1.4136E+00
BTU PER MIN	1.7573E+01	1.7573E-02	1.7572E-05	4.1965E+00	1.6667E-02	1.0000E+00	1.7576E+01	1.7576E+01	2.3560E-02
JOULES ABS PER SEC	9.9981E-01	9.9981E-04	9.9981E-07	2.3875E-01	9.4827E-04	5.6896E-02	1.0000E+00	1.0000E+00	1.3405E-03
WATTS (ABS)	9.9981E-01	9.9981E-04	9.9981E-07	2.3875E-01	9.4827E-04	5.6896E-02	1.0000E+00	1.0000E+00	1.3405E-03
HORSEPWR (ELECT)	7.4586E+02	7.4586E-01	7.4586E-04	1.7811E+02	7.0741E-01	4.2445E+01	7.4600E+02	7.4600E+02	1.0000E+00

CONVERSION FACTORS - ENERGY, WORK, HEAT

DESIRED UNITS:

GIVEN UNITS	JOULE	KG-METERS	FT-POUNDS	KW-HRS	HP-HRS (METRIC)	HP-HRS	LITRE-ATMOSPHERE	KILO CALORIES	BTU
JOULE	1.0000E+00	1.0197E-01	7.3746E-01	2.7778E-07	3.7764E-07	3.7251E-07	9.8687E-03	2.3885E-04	9.4787E-04
KG-METERS	9.8067E+00	1.0000E+00	7.2320E+00	2.7241E-06	3.7034E-06	3.6531E-06	9.6779E-02	2.3423E-03	9.2954E-03
FT-POUNDS	1.3560E+00	1.3827E-01	1.0000E+00	3.7667E-07	5.1208E-07	5.0512E-07	1.3382E-02	3.2388E-04	1.2853E-03
KW-HRS	3.6000E+06	3.6710E+05	2.6549E+06	1.0000E+00	1.3595E+00	1.3410E+00	3.5527E+04	8.5985E+02	3.4123E+03
HP-HRS (METRIC)	2.6480E+06	2.7002E+05	1.9528E+06	7.3556E-01	1.0000E+00	9.8640E-01	2.6132E+04	6.3246E+02	2.5100E+03
HP-HRS	2.6845E+06	2.7374E+05	1.9797E+06	7.4569E-01	1.0138E+00	1.0000E+00	2.6493E+04	6.4118E+02	2.5445E+03
LITRE-ATMOSPHERE	1.0133E+02	1.0333E+01	7.4727E+01	2.8147E-05	3.8267E-05	3.7746E-05	1.0000E+00	2.4202E-02	9.6047E-02
KILO CALORIES	4.1868E+03	4.2693E+02	3.0876E+03	1.1630E-03	1.5811E-03	1.5596E-03	4.1318E+01	1.0000E+00	3.9685E+00
BTU	1.0550E+03	1.0758E+02	7.7802E+02	2.9306E-04	3.9841E-04	3.9300E-04	1.0412E+01	2.5198E-01	1.0000E+00

CONVERSION FACTORS - ACCELERATION

DESIRED UNITS:

GIVEN UNITS	METERS PER SEC-SEC	CM PER SEC-SEC	FT PER SEC-SEC	FT PER MIN-MIN	KNOTS PER SEC	METERS PER HR-SEC	KM PER HR-SEC	MILES PER HR-SEC	FEET PER HR-SEC
METERS PER SEC-SEC	1.0000E+00	1.0000E-02	3.0479E-01	8.4667E-05	5.1440E-01	2.7778E-04	2.7778E-01	4.4703E-01	8.4667E-05
CM PER SEC-SEC	1.0000E+02	1.0000E+00	3.0479E+01	8.4667E-03	5.1440E+01	2.7778E-02	2.7778E+01	4.4703E+01	8.4667E-03
FT PER SEC-SEC	3.2810E+00	3.2810E-02	1.0000E+00	2.7779E-04	1.6878E+00	9.1139E-04	9.1139E-01	1.4667E+00	2.7779E-04
FT PER MIN-MIN	1.1811E+04	1.1811E+02	3.5998E+03	1.0000E+00	6.0756E+03	3.2808E+00	3.2808E+03	5.2798E+03	1.0000E+00
KNOTS PER SEC	1.9440E+00	1.9440E-02	5.9250E-01	1.6459E-04	1.0000E+00	5.4000E-04	5.4000E-01	8.6902E-01	1.6459E-04
METERS PER HR-SEC	3.6000E+03	3.6000E+01	1.0972E+03	3.0480E-01	1.8519E+03	1.0000E+00	1.0000E+03	1.6093E+03	3.0480E-01
KM PER HR-SEC	3.6000E+00	3.6000E-02	1.0972E+00	3.0480E-04	1.8519E+00	1.0000E-03	1.0000E+00	1.6093E+00	3.0480E-04
MILES PER HR-SEC	2.2370E+00	2.2370E-02	6.8180E-01	1.8940E-04	1.1507E+00	6.2139E-04	6.2139E-01	1.0000E+00	1.8940E-04
FEET PER HR-SEC	1.1811E+04	1.1811E+02	3.5998E+03	1.0000E+00	6.0756E+03	3.2808E+00	3.2808E+03	5.2798E+03	1.0000E+00

CONVERSION FACTORS - ENERGY PER UNIT AREA

DESIRED UNITS:

GIVEN UNITS	LANGLEY	CAL (15) PER SQ CM	BTU PER SQ FT	KW-HR PER CU METER	JOULES AB PER SQ CM
LANGLEY	1.0000E+00	1.0000E+00	3.6869E+06	1.1630E-02	4.1867E+00
CAL (15) PER SQ CM	1.0000E+00	1.0000E+00	3.6869E+06	1.1630E-02	4.1867E+00
BTU PER SQ FT	2.7123E-07	2.7123E-07	1.0000E+00	3.1544E-09	1.1356E-06
KW-HR PER CU METER	8.5985E+01	8.5985E+01	3.1702E+08	1.0000E+00	3.6000E+02
JOULES AB PER SQ CM	2.3885E-01	2.3885E-01	8.8062E+05	2.7778E-03	1.0000E+00

CONVERSION FACTORS - HEAT FLOW - POWER PER UNIT AREA (CAL ARE 15 DEG)

DESIRED UNITS:

GIVEN UNITS	WATTS PER SQ M	WATTS PER SQ CM	CAL PER SQ CM-SEC	CAL PER SQ CM-HR	BTU PER SQ FT-HR	BTU PER SQ FT-DAY	LANGLEY PER MIN
WATTS PER SQ M	1.0000E+00	1.0000E-04	2.3880E-05	6.6335E-09	3.1721E-01	1.3217E-02	1.4328E-03
WATTS PER SQ CM	1.0000E+04	1.0000E+00	2.3880E-01	6.6335E-05	3.1721E+03	1.3217E+02	1.4328E+01
CAL PER SQ CM-SEC	4.1876E+04	4.1876E+00	1.0000E+00	2.7778E-04	1.3284E+04	5.5348E+02	6.0000E+01
CAL PER SQ CM-HR	1.5075E+08	1.5075E+04	3.5999E+03	1.0000E+00	4.7819E+07	1.9925E+06	2.1599E+05
BTU PER SQ FT-HR	3.1525E+00	3.1525E-04	7.5281E-05	2.0912E-08	1.0000E+00	4.1666E-02	4.5169E-03
BTU PER SQ FT-DAY	7.5660E+01	7.5660E-03	1.8068E-03	5.0189E-07	2.4000E+01	1.0000E+00	1.0841E-01
LANGLEY PER MIN	6.9793E+02	6.9793E-02	1.6667E-02	4.6297E-06	2.2139E+02	9.2246E+00	1.0000E+00

INDEX

Accuracy, 2-13
Adiabatic lapse rate, 4-5
Aerosol, 2-39
Air density, 3-5
Air quality dispersion modeling, 1-7
Air temperature, 3-2, 3-5, 8-7
Alongwind dispersion, 4-21
Anemometer, 8-6
Anemometer height, 1-2
Annual concentration, 4-10
Area beneath Gaussian curve, 4-6
Area sources, 4-16
Areas within isopleths, 2-38
Assumptions, 2-38
Average concentration, 4-10, 4-11, 6-1
Average concentrations from
        instantaneous sources, 4-21
Averaging periods, 1-1, 2-8
Averaging time, 2-8, 2-11, 2-12, 5-1

Briggs-urban parameters, 4-12, 4-13,
        4-14, 4-15, 4-22, 8-27
Brookhaven dipersion parameters, 7-3
Building downwash, 7-2, 8-2
Buoyancy, 4-5
Buoyancy flux, 3-2, 8-1, 8-2, 8-3, 8-4,
        8-6, 8-8, 8-21, 8-23
Buoyancy-induced dispersion, 4-15,
        4-16, 7-2, 8-9
Buoyant rise, 3-1, 3-2, 4-12, 4-15
Buoyant turbulence, 1-4, 1-5, 2-7, 4-13

Centerline, 2-2, 4-17
Clear, 8-26, 8-29, 8-30
Cloud cover, 2-7, 8-3, 8-7, 8-8
Complete eddy reflection, 2-39
Concentration, 2-4, 2-15, 4-1
Concentration at plume centerline, 8-18
Concentration above ground, 8-15, 8-17
Concentration distribution, 2-1, 4-5,
        4-20
Concentration level-of-concern, 2-15
Concentration, mass per volume, 4-1
Concentration, volume per volume, 4-1
Coning, 1-6, 1-7
Conservation of mass, 2-1
Continuous emission, 2-38

Convective, 2-12, 5-1, 5-2
Convective scaling parameter, 3-5
Coordinate system, 2-2, 4-12
Coordinates, 1-7, 6-1, 7-3, 8-6
Critical wind speed, 4-12, 8-20
Crosswind distance, 4-18, 6-1, 7-3, 8-5,
        8-13, 8-27
Crosswind distribution, 2-2
Crosswind-integrated concentration, 4-8
Crosswind-integrated groundlevel
        dosage, 8-24
Crosswind spreading, 2-5

Diameter, 8-7
Dilution, 1-2, 4-12, 5-2
Direction of transport, 1-1
Diskette, 6-1, 7-1, 7-2
Dispersion, 1-4, 4-1, 4-2, 4-20, 6-1, 7-1
Dispersion model, 1-7, 6-1
Dispersion parameters, 2-8, 2-12, 2-15,
        8-27, 8-31
Distance, 2-15, 8-19, 8-20
Distance to maximum groundlevel
        concentration, 2-12, 2-13, 2-14,
        4-7, 8-1, 8-2, 8-3, 8-4, 8-5, 8-6,
        8-8, 8-9, 8-11
Distance to final rise, 3-4
Distance to a particular value of
        concentration, 2-15
Downwind distance, 2-5, 2-12, 2-15,
        2-17, 2-18, 4-4, 4-10, 4-16,
        4-18, 4-19, 6-1, 7-3, 8-2, 8-7,
        8-9, 8-10, 8-12, 8-13, 8-22,
        8-30, 8-31, 8-32
Dry adiabatic, 1-5

Eddies, 1-4, 1-5, 4-15
Eddy diffusivity, 4-5
Eddy reflection, 2-1, 2-39, 8-27
Edge effects, 4-18
Effective height, 2-4, 2-13, 2-14, 2-18,
        2-27, 3-1, 3-4, 4-3, 4-10, 4-11,
        4-17, 6-1, 7-2, 7-3, 8-1, 8-2, 8-3,
        8-4, 8-6, 8-8, 8-9, 8-11, 8-18,
        8-19, 8-21, 8-23, 8-25, 8-31
Elevated continuous sources, 4-7
Elevation angle of the sun, 7-4

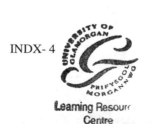

Learning Resource
Centre